Letting God Create Your Day

Volume 5

Paul A. Bartz

Creation Moments™

**Letting God Create Your Day: Volume 5
by Paul A. Bartz**

Copyright © 2004 Creation Moments, Inc.

Creation Moments, Inc.
P.O. Box 260 Zimmerman, Minnesota 55398
www.creationmoments.com
800-422-4253

Cover Photo: Our Creator created the myriad forms of life, each by His Word "according to their kind"(Genesis 1). Indeed, modern genetic science as well as the fossil record reveals that the variations of species do not go beyond their "kind", being limited by the particular genetic compatibility and diversity within breeding pairs. Hence, chickens remain chickens, though there are many varieties, and people remain people, though there are different colorations. But the variation is limited by the original genetic content of breeding pairs within the species kind, contrary to the hypothesis of evolutionism.
(Photo by Kathleen Cadwallader featuring her littlest brother Russell Cadwallader, children of Creation Moments Board Chairman Mark Cadwallader.)

ISBN 1-882510-09-7

Printed in the United States of America
Printing and production costs for this book were partially underwritten by friends and supporters of Creation Moments.

Foreword

Some 25 years ago when I became seriously involved in creation ministry polls pretty consistently showed that over 50 percent of the population accepted some form of evolution. Those who accepted some form of creation were usually in the upper 30 percentile. The rest were atheists or offered the "I don't know" answer. Many Christians accepted some form of evolution because they thought that they had no intelligent alternative in the Bible's teaching about creation.

Young people who had grown up in the church were leaving because what they had learned in Sunday school about creation didn't harmonize with the evolution they were getting in school. Many felt that if what they had learned about creation was wrong, what about what else they had learned. Perhaps, they thought, it was also true, as some said, that Christ didn't rise from the dead. The entire basis of their salvation had been undermined.

I know this because I have, over the years, talked with thousands of these young people.

Compared to today, there were relatively few voices speaking out in favor of creation. Very little original scientific research had been done by believing scientists. There were only a limited number of resources on creation for churches and families to use in support of the Bible's account of creation.

Millions of words later, things have really changed for the better. There is a wealth of materials for Sunday schools, homeschoolers and families. The youngest children can color in their dinosaur coloring books, and even the most advanced scientists can find original creationist research.

All of this is having an effect. In the last few years polls are showing that 50 (sometimes even a little more), percent of our population now believe in some form of creation, with those accepting some form of evolution is now in the upper 30 percentile! There's still much work to be done. That's why we're pleased to offer you *Letting God Create Your Day, Volume 5*. We at Creation Moments Ministries are pleased to have been blessed to do our part in getting out the truth about creation over all these years. We pray that you would enjoy growing in your wonder of the creation, and grow in faith through this volume.

Paul A. Bartz
Author, Creation Moments™

Has a Solar System Like Ours Been Discovered?

Psalm 8:3-4b
"When I consider thy heavens, the work of thy fingers, the moon and the stars, which thou hast ordained, what is man that thou are mindful of him...?"

We have all heard reports of astronomers claiming that they have discovered a planet orbiting some nearby star. Each announcement results in speculation about whether there is intelligent life on the newly discovered planet. Lost is the fact that some planets are larger than Jupiter and made of gas. Scientists discover these planets by studying hundreds of telescopic pictures of a suspect star to see if there may be a tiny wobble in its position. A wobble means the star has at least one orbiting planet, meaning that this solar system does not have multiple planets like ours does.

In April 1999, two teams of researchers announced they had discovered three planets orbiting nearby star Upsilon Andromedae. But, this solar system is not like ours either. The inner planet is so close to the star that it is hotter than Mercury—no life there. The second planet's orbit takes it close to the star then swings wide away from it. Such an orbit would provide temperature conditions far too erratic for life like that on Earth. The third planet is too far from Upsilon Andromedae and thus too cold to provide a home for life like ours.

Our solar system is not a result of chance but has been uniquely designed for life by a loving Creator. Not only is Earth's distance from the sun just right for life, but its orbit provides the even, moderate conditions that life requires.

Prayer: Thank you, dear Father in heaven, for preparing a creation for us in which every detail supports us and glorifies you. In Jesus' Name. Amen.

REF: R.Cowen, "Astronomers Find Planetary System." *Science News,* v. 155, April 17, 1999. p. 244

Sin Stinks!

Genesis 1:31
"And God saw everything that He had made, and behold it was very good. And the evening and the morning were the sixth day."

New findings about how the human nose works may help explain why a world that was perfect when it was made has come to have so many stinky things in it. Humans and mammals hold the record among all creatures, with about 1,000 genes for odor receptors. But with only 1,000 different kinds of odor receptors, how is it we can smell over 10,000 different kinds of scents?

For example, nonanoic acid uses the same receptors as nonanol, plus two others. Nonanol literally produces the smell of a rose. Yet those same receptors, plus the extra two that nonanoic acid activate, will produce a cheesy odor. Even more amazing is the discovery that virtually identical molecules can produce very different scents. Octanol produces a rosy, orangy scent, when octanoic acid produces a rancid, sweaty odor. Yet the two molecules are identical except for an extra side chain of atoms.

This may go a long way in explaining how a perfect creation of vegetarian creatures came to have rotten smells and to be meat eaters. A minor modification of a "good" scent can turn it into a "bad" scent. This seems to be confirmed by the discovery that the same receptors that allow mice to smell grain and seeds allow rats to smell meat. The decay man's sin brought to the creation didn't mean that God had to make new things. "Good" molecules simply decayed into a slightly different form with "bad" effects. We thank God we have a Savior in His Son Jesus Christ Who will deliver us from the decay that is in the world!

Prayer: Dear Father in heaven, we thank you that in Jesus Christ, you have delivered us from the decay in the world, which is a result of our own sin. Amen

Ref: John Travis, "Making Sense of Scents," *Science News,* v. 155, April 10, 1999, p. 236

Is Marriage Just a Piece of Paper?

Matthew 19:4-5
"And He answered and said unto them, 'Have ye not read that He which made them at the beginning made them male and female. And said, For this cause a man shall leave his father and mother and shall cleave to his wife, and the twain shall be one flesh?'"

With the creation of Adam and Eve, God also created the marriage relationship and blessed it. God knows how we are made. He knows how we work. And He had all that in mind when He designed the marriage relationship. For this reason, one would expect that trying to fashion a family by simply living together would result in less happiness than when properly married.

Numerous studies in recent years have shown the results of couples living together without the benefit of marriage. One 1992 study, conducted at Rutgers University, showed that even cohabitors who eventually got married had a 46 percent greater chance of divorce than those who married before living together. A more recent study by a Bowling Green University researcher measured and compared the depression scores between married couples and cohabitors. Married people had an average depression score of 13.3, while cohabitors showed a much higher depression score of 17.2.

Cohabiting is even worse for the children. Thirty-six percent of the children born to cohabitors will see their parents divorce. Only one-third of the children born to married people will see their parents divorce. It is clear that in His love for us, God designed marriage for our happiness and blessing.

Prayer: We thank you, dear Father, for the gift of marriage. Bless and strengthen marriage among us. In Jesus' Name. Amen.

Ref: Mona Charen, "Before you decide on moving in," *Washington Times,* March 26, 1999, Third Age News Service,
"Marriage gets the nod for satisfaction," *Brainerd Daily Dispatch*, (Minn.) September 6, 1998.

Science is Catching Up with Genesis

Genesis 8:22
"While the earth remaineth, seedtime and harvest, and cold and heat, and summer and winter, and day and night shall not cease."

Climate researchers are developing more complex computer models in an effort to discover the effects of climate changes. One unexpected conclusion is that greenhouse warming of the Earth may not be happening. And if it is, it may not be a bad thing. Researchers have also had to admit that the natural climate changes of the earth are much wider than originally thought.

In one computer model, researchers assumed that carbon dioxide would increase in the atmosphere by one percent a year. In another they assumed one-half of one-percent increase, and in yet another model they assumed no increase. The study yielded mixed results. Central and western Europe would see no abnormal change in river runoff. River runoff would increase in northern Europe, but decrease in southern Europe.

Their model also showed that wheat crops would increase in several European countries, with no abnormal changes in other countries. In fact, researchers found that an increase in carbon dioxide would cause a 9 to 39 percent increase in wheat crops throughout Europe.

Science is discovering that what God promised Noah after the Flood is true: "While the earth remains, seedtime and harvest, cold and heat, winter and summer, day and night shall not cease."

Prayer: We thank you, Lord, that you have given us a beautiful and durable Earth to support our needs. Make us good stewards of it, rejoicing without fear of the future. Amen.

Ref: C. Wu, "Fickle climate thwarts future forecasts," *Science News,* v. 155, February 27, 1999, p. 133

The Flower that Favors Bumblebees

Matthew 6:28-29
"'And why take ye thought for raiment? Consider the lilies of the field, how they grow: they toil not neither do they spin: And yet I say unto you that even Solomon in all his glory was not arrayed like one of these.'"

The tropical-looking flower Virginia Meadow Beauty offers some rare tricks for those who would pollinate it. A honeybee can poke around the flower all day and never get any pollen from the gaudy pink flowers. In fact, researchers report that honeybees don't seem to know what to do with the flowers. Rather, the Virginia Meadow Beauty is pollinated by bumblebees, not honey bees.

The pollen of a newly opened pink flower is good only for that day. But the flowers stay open for several days. It signals the bumblebees that the pollen in old flowers is not good by changing colors. Yet the spent flowers are serving one more purpose. Researchers report that large displays, even if mostly spent flowers, attract more bumblebees than small displays of all fresh flowers.

How is it the bumblebee can get the pollen and the honeybees can't? They just buzz. And when a bumblebee buzzes, the flower ejects its pollen at 30 times the force of gravity—a force greater than any astronaut must endure! It's not that the bumblebee gets the pollen because it is larger than a honeybee—even bumble bees no larger than the honeybee's head can get the pollen. It's the frequency of buzzing. The Lord has abundantly provided for the Meadow Beauty. And He has provided for your forgiveness and salvation in His Son, Jesus Christ.

Prayer: I thank you, Lord that you so generously provide for all your creatures. Increase my faith that you will also provide me with all I need, beginning with salvation. Amen.

Ref: J. Travis, "Color code tells bumblebees where to buzz," *Science News*, v. 155, April 3, 1999. p. 215

Did King David Go to the Mall?

1 Kings 20:34
"And Ben-hadad said unto him, 'The cities which my father took from thy father I will restore; and thou shalt make street for thee in Damascus, as my father made in Samaria.' Then said Ahab, 'I will send thee away with this covenant.'"

Did King David go to the mall? Malls or shopping areas are hugely popular today. Archaeologists are now asking whether shopping malls originated in the 11th century BC.

Archaeologists have long speculated on the purpose of the buildings they call tripartite-pillared buildings. These were long, rectangular buildings that were divided into thirds by two interior rows of pillars. Thirty-five of these buildings have been found at 12 sites. Some thought they were storehouses. Others suggested they were stables.

Their size and the thickness of the walls made it clear to archaeologists that these were not private homes. Nor has a used household bowl ever been found in one. But many new, unused vessels, including lamps, which have no soot, have been found in them. In addition, evidence of high windows has been found at one site. Based on this and other evidence, one archaeologist has suggested that these buildings were, in fact, marketplaces such as those described in 1 Kings 20:34. They originated in the land that Ben-hadad ruled.

David could have gone to the mall, for the Philistines had them. And King Ahab appears to have been active in getting malls built in Judah. Perhaps those people three thousand years ago were not so different from us today.

Prayer: Dear Father, we thank you that we share the same humanity as all since Adam, rather than having descended from apes. Amen.

Ref: Moshe Kochavi, "Divided Structures Divide Scholars," *Biblical Archaeology Review*, May/June 1999, p. 44

Lizard Language

Acts 2:6
*"Now when this was noised abroad, the multitude came together,
and were confounded, because that everyman heard them speak
in his own language."*

Since God is the Author of all language, we shouldn't be
surprised to find many and varied forms of communication among the
creatures He made.

The recent discovery that even lowly lizards communicate with
one another has offered several surprises for those who believe we
evolved. After analyzing over 1,500 displays by sagebrush lizards,
researchers say that there is no question that the reptiles are
communicating with one another. Lizards communicate through complex
body language that follows recognizable rules of grammar. It is made up
of three parts, which include head bobbing, "push ups," and leg motion.
Mixing and matching one or more of these actions results in over six
thousand possibilities!

But what do lizards say to one another? Repeated observations
have convinced researchers that they can interpret what the lizards are
saying to each other. Much of the communication is used to woo a mate.
Lizards will also brag to other lizards when they find a particularly good
rock. But a lizard may also warn away another lizard who might want to
share his rock.

Not only is such complex communication among lowly lizards
not expected by evolution, but also researchers noted that the pattern of
communication over separated populations does not follow expected
evolutionary patterns. Yes, even the lowly lizard glorifies its Creator!

*Prayer: We thank you, dear Father, for the gift of language.
We especially thank you for your trustworthy Word. Amen.*

Ref: Susan Milius, "When Lizards Do Push-Ups," *Science News*, v. 155, February 27, 1999, p. 142

Bacteria the Size of a Whale?

Genesis 1:20

"And God said, 'Let the waters bring forth abundantly the moving creature that hath life, and fowl that may fly above the earth in the open firmament of heaven.'"

The abundance and variety of living things that God made continues to amaze scientists. All of us learned in school that bacteria are too small to see without a microscope. So when scientists studied a newly discovered type of bacteria that is visible to the naked eye, they couldn't believe it was bacteria! More study convinced them that the sulfur pearl of Namibia is indeed the largest known bacteria.

The sulfur pearl of Namibia was discovered in deep-sea sediments in 1997. It is part of a strange family of bacteria that generate energy for life by oxidizing sulfur. Such bacteria had been found earlier in sediments found off the South American coast, but they are nowhere near the size of the sulfur pearl. Not only does the sulfur pearl need no light to live, but also it is an excellent reflector of light, which makes it easy to see. The sulfur pearl is so large that smaller bacteria colonize its outer sheath. If the average bacteria were the size of a mouse, the sulfur pearl would be the size of a great blue whale, the largest animal on earth! It is in fact about as big as a pinhead.

Despite its great size, the sulfur pearl has no more cytoplasm than the average bacteria. About 98 percent of the bacterium's interior is taken up by a fluid-filled sack in which the bacterium stores nitrates, which it uses to metabolize sulfur. This allows it to go for months without finding new nitrates. God has generously creatively provided for all His creatures, even this bacterium.

Prayer: Dear Father, we thank you for the variety and creativity with which you have provided for all your creatures. Amen.

Ref: S. Milius, "Digging bait worms reduces birds food," *Science News*, April 17, 1999, v. 155, p. 246

How Important Are Fathers?

Luke 1:17
"And he shall go before Him in the spirit and power of Elias, 'to turn the hearts of the fathers to the children,' and the disobedient to the wisdom of the just, to make ready a people prepared for the Lord."

The role of father was given to us by God. But over the last several decades, some have tried to redefine or redesign the family in a way that makes fathers unimportant or unnecessary. Numerous studies have shown what happens when we do this.

•Between 1960 and 1990 the percentage of children living apart from their fathers doubled to 36 percent.
•Studies show that it is much worse for a child to lose a father through divorce than through death.
•Fatherlessness has been shown to be a contributing factor to early sexual activity.
•Teen suicides, shown to increase in fatherless families, have tripled since 1960.
•Scholastic assessment test scores have dropped 75 points since 1960; the drop is linked to the lack of fathers at home.
•Teenage boys raised without a father are more likely to get in trouble with the law.
•Children without their fathers are much more likely to suffer child abuse, according to several studies.
•One 26-year-long study found that the most important childhood factor in developing empathy is a father's involvement in the family.

Yet another study has found that 90 percent of the children who go to church with their parents will remain active in church through their teens. If neither parent comes with them, only 40 percent remain faithful. If only dad comes with them, 80 percent remain faithful. We must admit that we cannot improve on God's design for the family.

Prayer: Father, thank you for fathers, and give us more faithful fathers. Amen.

Ref: David Popenoe, "Life Without Father," *Readers Digest*, February 1997, p. 65, *Lutheran Witness*, June 1995 v. 147, p. 7

A Simian Shakespeare Theatre?

Romans 1:20
"For the invisible things of Him from the creation of the world are clearly seen, being understood by the things that are made, even His eternal power and Godhead, so that they are without excuse:"

You may never see a troop of monkeys traveling from city to city to perform Shakespeare, but some researchers now believe that monkeys do indeed develop culture. This unexpected discovery does not support evolution. Rather, it shows how the creation itself reflects the nature of our Creator.

God's very act of creation was an application of His knowledge to impress order on the creation. But human beings are not the only creatures who use knowledge to increase order and make life easier. When such knowledge and order is passed on to the next generation, you have what we call culture.

Researchers studied the scientific literature on four populations of African chimps. Their study revealed that these chimp populations do indeed pass knowledge unto their children. For example, one population eats soldier ants by holding a stick near the entrance to the nest. Once the stick is filled with ants, they sweep the stick clean with their lips. At another preserve, the chimps collect the ants the same way, but then sweep the stick clean with their fingers. Researchers have even seen mother chimps teaching their children certain skills. They have seen mother chimps leave nuts and stone "hammers" lying around to teach their children how to open nuts. One mother even did a slow-motion demonstration of nut-cracking for her child.

It's not the ability to pass on information that makes us unique. It's God's special love for us in sending His Son to save us!

Prayer: Dear Father, we thank you for your love for us in making yourself evident in the creation. Help us witness your love, too. Amen.

Ref: B. Bower, "Chimps may put their own spin on culture," *Science News,* v.154, December 12, 1998, p. 374

14

The "Strange Berry"

Psalm 71:17
"O God, Thou hast taught me from my youth; and hitherto I have declared thy wondrous works."

Sometimes it seems as if God made some creatures just to show us He could do the impossible. Many of these creatures, by their strange nature, offer direct challenges to evolutionary theory, since there was no evolutionary need for their unique nature.

One such creature is a bacterium that has been labeled "the toughest bug on earth." Its Latin name means "strange berry that withstands radiation." It can withstand thousands of times the amount of radiation that would kill a human. The bacterium was first isolated in the 1950s, but a scientist who began studying the bug in 1988 said, "I had difficulty believing anything like this could exist."

Many bacteria form hard capsules around themselves in response to radiation. While this provides some protection, the "strange berry" doesn't form a capsule, and still survives better than any other bug. While 500 to 1,000 rads of radiation would kill a human being, the "strange berry" can withstand 1.5 million rads. The radiation shatters the "berry's" DNA into hundreds of fragments, a hundred times the fragmentation that is fatal to other bacteria. But a couple of hours later, the DNA is stitched back together, free of all mutations.

Evolutionists are puzzled because there is no environment containing this much radiation. Why would evolution develop such a creature? The "strange berry" not only challenges evolution directly, but also declares the skill and wisdom of its Creator!

Prayer: We praise you, dear Father, for your wondrous works, which challenge unbelief and declare your glory. Amen.

Ref: John Travis, "Meet the Superbug, "*Science News,* December 1998, v.154, p.376

A Noisy, Bird-Brained Harem

Job 28:20-21

"'Whence then cometh wisdom? and where is the place of understanding? Seeing it is hid from the eyes of all living, and kept close from the fowls of the air.'"

Scripture frequently makes reference to the fact that birds are not very smart compared to human beings. But the tropical wetland bird called the jacana shows that you don't have to be very smart to be deceptive.

The jacana is one of only 20 species of birds in the world where the female leaves the care of the young to the males. When researchers studied a flock of jacanas, they observed that the males staked out their territory on floating vegetation, often getting into violent fights with other males. Then the females, which are about 60 percent larger than the males, fought with each other for exclusive rights to up to four male territories. Once territories were established, the females would visit each of the males in her territory, mating with each.

Once the eggs are laid in each male's nest, the female shows no more interest in her offspring. The male will care for the eggs and hatchlings, until the time the youngsters are ready to leave home. But the smaller males have their own strategy for dealing with their situation. They yell. Researchers say a yelling male is really blackmailing the larger female into giving him some attention. A yelling male attracts the attention of other nearby females who might want to take him into her own harem, so his mate comes running to pay attention to him. Sometimes males will even fake an emergency, which brings his mate in a hurry!

God is the source of all wisdom, and He gave each of His creatures enough wisdom to conduct their lives.

Prayer: Father, I thank you for your wisdom in Holy Scripture. Grant me understanding as I read Your Word. Amen.

Ref: S. Milius, *Science News*, March 6, 1999, v. 155, p.149

This Flower Is a Real Stinker

Isaiah 40:28
"Hast thou not known? Hast thou not heard, that the everlasting God, the Lord, the Creator of the ends of the earth, fainteth not, neither is weary? there is no searching of His understanding."

Dung beetles may not be a very pleasant subject, but after all, someone has to do the housekeeping. So God created what man has classified into 30,000 species of dung beetle. Many will form their finds into a ball about the size of a croquet ball and roll it home for the wife and kids. Others tunnel beneath a find and set up housekeeping there.

A purple flower in Borneo, however, cuts into the productivity of one of the tunneling species without giving it anything in return. This flower smells, not like a rose, but like dung. It relies primarily on this dung beetle for pollination. Studies show that most other pollinators will examine the flower, but, perhaps put off by the smell, won't attempt to pollinate it. But, the dung beetle is a successful pollinator as it searches the flower for food. The flower's unpleasant odor seems designed to do nothing other than fool the beetle into pollinating it. It apparently offers the beetle nothing.

It would be nonsense to say that this flower designed this strategy to attract a beetle able to pollinate it. It would be just as silly to say that the impersonal forces of evolution designed this deceptive plant. Only an all-powerful, all-wise Creator could have designed this relationship, which intricately weaves together needs and abilities to make the whole creation work!

Prayer: Dear Father in heaven, give us more wisdom and understanding, that we may be more like you. In Jesus' Name. Amen.

Ref: S. Milius, "Stinking beauty betrays dung beetles," *Science News*, January 23, 1999 v.155, p.55

Has Your Great-Great-Grandfather been Born Yet?

Genesis 5:2
*"Male and female created he them, and blessed them, and called
their name Adam, in the day when they were created."*

Has your great-great-grandfather been born yet? If he hasn't, he's
not your great-great- grandfather! This might seem like a silly question.
But when it comes to the claimed evolutionary fossil evidence for man's
evolution, our claimed great-great-grandfather has turned up in the
wrong period of history. A re-dating of *Homo erectus* fossil skulls
discovered in 1931 and 1933 has placed these supposed "human
evolutionary ancestors" squarely in human history. This dating is
supported by the geological setting in which they were found. In fact,
two excavation sites right next to each other together yielded both
modern human and *Homo erectus* remains!

The most likely age for the *Homo erectus* remains were 27,000
years before present. You must remember that these are years according
to the evolutionary interpretation. But what it means is that *Homo
erectus* and humans like us lived at the same time. This is troubling to
evolutionary scientists, who have tried several tactics to avoid the
problem. First, they argued that the human fossils were not in the same
geological layer as the *Homo erectus* remains. But photographs
contradict this claim. Then, they argued that the human remains must
have washed into the area from higher ground. But the good condition of
the human remains and related bones ruled out that possibility.

In other words, *Homo erectus* offers no support for human
evolution, another attempt to contradict God's Word about our special
creation falls to failure!

***Prayer: Dear Lord, we thank you that you have not only
specially created us, but also that you died to save us. Amen.***

Ref: Marvin Lubenow, "Alleged Evolutionary Ancesors Coexisted With Modern Humans," ICR,
Impact No. 286, April 1997

Not So Bird-Brained

Genesis 1:21
"And God created . . . every winged fowl after his kind; and God saw that it was good."

How do you take a much-needed nap or get a good night's sleep when you must be alert to danger? Human beings designate people to stay awake and watch for danger when they sleep. Then, they set up warning systems.

Since birds are much more vulnerable to danger, God gave them the ability to do these things on a simple scale, and then He gave them one more amazing gift. Many birds are able to put half of their brain to sleep, while the other half stays awake and alert. The eye that serves the sleeping half even closes, while the eye that serves the waking half of the brain stays alert, searching for danger. After the first sleeping half is nicely rested, it wakes up, and the other half goes to sleep. Of course, these birds can also put both sides of the brain to sleep at the same time, just as we do. Whales, seals and dolphins are among the non-birds that can half-brain sleep as well.

Evidence of the clever design of this half-brain sleep ability can be seen in the fact that where two birds sleep next to each other, each bird will put the half of the brain that faces its partner to sleep. Once that half is rested, they change places, so the other half of their brain can sleep.

Scripture tells us that God looked at His creation and declared it "good," and God's standard for good is perfection. His ingenious design, which would allow birds to protect themselves even when sin and death entered the creation, is part of the goodness of God's creation.

Prayer: I thank you Lord that you made a creation perfect and so beautiful that even sin could not completely destroy it. Amen.

Ref: S. Milius, "Half-asleep birds choose which half dozes," *Science News*, February 6, 1999, v.155, p.86

Unnatural Selection?

Luke 12:33
"Sell that ye have and give alms; provide yourselves bags which was not old, a treasure in the heavens that faileth not, where no thief approacheth neither moth corrupteth."

Just about every public school textbook includes the example of the peppered moth. The moth is used as a prime example of natural selection. Supposedly, as the trees in the English countryside began to be covered with coal pollution in the mid-1800s, the light-colored tree trunks became darker. The peppered moth exists in two varieties, black and white, and was said to rest on the tree trunks. As the trunks turned black the white variety was more easily seen and picked off by the birds, leaving the black variety to multiply. Students were told this was natural selection.

But since the 1980s, numerous studies of moth populations and how they live have called into question the use of the peppered moth as an example of evolution. Several additional population studies in polluted and unpolluted forests show little correlation between whether there are lighter or darker moths. After pollution control laws went into effect in England, the population of dark moths decreased in the north, but increased in the south! In addition, the nocturnal moths do not rest on the tree trunks during the day, but stay hidden under the branches higher in the tree. In the famous photograph shown in every biology textbook— of a pair of moths resting on a tree, the moths were actually glued to the tree trunks to provide the picture!

Let's not give away the treasures of the truth of God's inerrant Word for the false treasure of earthly theories which deny our Creator.

Prayer: Thank you, Father that your saving Word is trustworthy. Amen.

Ref: Jonathan Wells, Ph.D., "Second Thoughts about Peppered Moths," April 6, 1999

One Smelly Amoeba

Job 9:25-26
"Now my days are swifter than a post; they flee away, they see no good. They are passed away as the swift ships, as the eagle that hasteth to the prey."

Not all dangerous predators can be seen. One of the most dangerous predators in a drop of pond water is *Amoeba proteus*. This amoeba literally terrorizes its one-celled pond-mates, who it can smell the amoeba coming to engulf them. Among the amoeba's favorite snacks are creatures of the genus *Euplotes*. But *Euplotes* can defend themselves. When some members of this genus smell the amoeba coming, they dart away to safety. Others can grow pikes to defend themselves. Some can even grow hard shells to protect themselves.

This scenario leaves a problem for evolution. What benefit to a predator is there in warning its prey? Evolution should have selected against the amoeba's ability to produce its unique scent, called the A-factor. Now a new discovery has revealed the amoeba's scent is really a clever design, pointing to a Designer. These amoebas reproduce by splitting, so they are surrounded by clones. The A-factor is just strong enough to allow the *proteus* to identify clone-mates so that they don't eat each other, but not so strong that it prevents *proteus* from getting enough to eat.

God sent His Son, Jesus Christ, to save us from another unseen predator, the devil. The evil in the world is the warning scent of the devil's presence. God has placed many evidences within the creation that He is the Creator so that as our relatively few days on this earth flee by, we will be drawn to His Son, and, believing in Him, receive forgiveness and salvation from sin, death and the devil.

Prayer: We thank, you, dear Father, that in your love, you sent Your Son to save me from sin, death and the devil. Amen.

Ref: S. Milius, "Amoeba betrayed by anticannibal defense," *Science News*, March 20, 1999, v. 155, p. 182

The Importance of Good Parents

Ephesians 6:1
"Children, obey your parents in the Lord; for this is right."

Do good parents undo their best efforts when they place a child in day care? Who has more influence on a child's development when that child must be placed in day care? Recently we have heard about many studies, some of which arrived at conflicting conclusions. Some studies said that parents have little or no effect on their children's development. Others concluded that day care doesn't influence a child's development at all. Now, unique new research seeks to clear this up.

The new two-pronged research effort is unique because it relied on direct observation of children and caregivers. The first prong of the study, funded by the National Institute of Child Health and Human Development, observed 1- to 3-year-olds in over 1,300 families. These families ranged from stay-at-home moms to those who used day care. Home backgrounds were studied, mothers were videotaped, and children were tested for language development and similar skills. This part of the study concluded that mothers who are warmly involved with their children had the greater influence on their child's development. They also concluded that average children who go to day care and have an involved, caring mother, will be stunted by poor day care, but modestly helped by good day care.

The second prong of the study rated day care in nine states, and then extended its results to the entire country. They rated 61 percent of day care program as "fair" or "poor, " with 39 percent rated as "good" or "excellent." What's clear from these results is that in God's order of things, nothing can replace good parenting!

Prayer: Lord, help me to be a good parent, and to be good to my parents. Amen.

Ref: S.M. "Good parents still make the difference, "*Science News*, February 6, 1999, v.155, p.91

Glow, Little Octopus

John 8:12
"Then Jesus spake again unto them, saying, 'I am the light of the world. He that follows me shall not walk in darkness, but have the light of life.'"

The deep, dark depths of the ocean are filled with many varieties of luminescent creatures. But among the octopus, only two or three species have the ability to luminescence. But that has all changed with a remarkable series of discoveries about an already known species of octopus.

The red octopus, *Stauroteuthis syrtensis*, lives in the deep waters off the east coast of the United States. While examining a foot-long specimen of the red octopus in 1997, scientists got a surprise when they turned off the lights in the lab, the scientists were amazed to see the octopus's suckers glowing. The blue-green glow, they discovered, glows brightest at 470 nanometers, a frequency that travels well under water. Scientists say that this glow might explain how the octopus makes its living. It doesn't eat what most octopods eat. Rather, it eats tiny crustaceans whose shallow water cousins are drawn to light. If the deep-water versions are drawn to light, all this clever little octopus needs to do to eat is turn on the porch light.

The frequency of the light and the unusual diet of this octopus all provide evidence of God's all-wise design. If these special features had depended on chance to develop, this octopus would not exist.

God is the Creator and Source of all light. The perfect light of His truth in the Gospel shines at just the right frequency to draw us to the forgiveness of sins and salvation found only in Jesus Christ.

Prayer: Father in heaven, we thank you for the light of the Gospel which brings us to the light of the life Christ has won for us. Amen.

Ref: S. Milius, "Octopus suckers glow in the deep, dark sea," *Science News*, March 13, 1999, v. 155, p.167

"Too Many Notes"

1 Chronicles 13:8
"And David and all Israel played before God with all their might, and with singing, and with harps, and with psalteries, and with timbrels, and with cymbals, and with trumpets."

Perhaps you remember the line from the film *Amadeus* in which someone observes of Mozart's music, "Too many notes." That statement reflects some peoples' opinion of Mozart's highly complex music. But researchers are finding that this mathematical complexity may be the reason for what some call "the Mozart effect."

Researchers at several universities around the United States have confirmed that Mozart's complex music has positive effects for both adults and children. In one study, rats were subjected to Mozart for 12 hours a day, beginning 4 weeks before birth. A second group heard only silence, a third heard only a constant hissing sound, while a fourth group heard only minimalist composer Philip Glass. When the rats were old enough to run a maze, they were tested. The Mozart rats not only ran the maze considerably faster than any of the others, but they made fewer mistakes. Other research has shown that adults do better on intelligence tests after hearing Mozart. While this effect is temporary in adults, children exposed to Mozart show a permanent improvement.

It is thought that the complex nature of Mozart's music encourages the brain to make more connections within itself. The more connections you have, the smarter you are. Surely music is a gift of God that benefits us and could never exist in a universe created by the forces of chance.

Prayer: Thank you, Lord for the gift of music. Help me to use music to praise you and to tell others of your salvation. Amen.

Ref: John Fauber, "Mozart is music to the brain's ears," *The Christian News*, January 18, 1999, p.9

American Flashers

Matthew 5:16
"'Let your light so shine before men, that they may see your good works and glorify your Father which is in heaven.'"

Some time ago, **Creation Moments** aired a program about Asian fireflies that flash in unison. The unspoken question was, "Aren't there some American species that flash in unison?" Someone in Tennessee who saw the same reports on which that our program was based had the answer to that question. She contacted a researcher at Georgia Southern University to report an American species that clearly flashes in unison. At first, researchers had a difficult time believing that there were other firefly species that could accomplish this difficult task. But now several American species that flash in unison have been identified.

The newly identified species is found in the eastern United States, as far north as Pennsylvania and all the way down to Georgia. Another species has been found in Texas. Yet another species is found in the Gulf States. In each case, the males synchronize their flashes, creating a dramatic light show. Some species flash in unison for hundreds of flashes, producing shows that last up to three or four minutes.

Researchers have tried to explain this complex behavior in what is supposed to be a simple species. What troubles them is that every explanation they come up with credits the firefly with having some sort of pacemaker in its tiny brain. To those with an evolutionary view, this is unacceptably complex. But for those of us who know we have an all-wise Creator, the synchronized flashing of fireflies is but another sign of the excellence of His work.

Prayer: Father, let my light shine before men so that others may see your work. Amen.

Ref: Susan Milius, "U.S. Fireflies Flashing in Unison," *Science News*, March 13, 1999, v. 155, p. 168

Doctors Study GOD

John 11:11-12
"These things said He, and after that he said to them, 'Our friend Lazarus sleepeth, but I go, that I may awake him out of sleep.' Then said his disciples, 'Lord, if he sleeps he shall do well.'"

Physicians have been mystified by and are in the process of studying GOD. But in this case, the acronym "GOD" stands for "generation of diversity." The diversity that amazes them is the diversity found in our immune system. While we have fewer than 100,000 genes, our immune system is able to generate tens of millions of distinct antibodies to defend and protect itself from disease. How can only 100,000 genes produce thousands of times that many antibodies?

Actually, we have two immune systems. The first is called the innate system. This rapid response system and relies on common characteristics found in most microbes to bind and conquer them. The second immune system is called the combinatorial immune system. It is this second immune system that generates tens of millions of different and very specific antibodies. Its response is slower than the innate system because it responds to an infection by making and sending a variety of antibodies at invaders. Depending on which antibodies recognize the invader, the combinatorial system then begins to manufacture more specific antibodies until the invader is conquered.

Researchers studying the details of the immune system are astonished at how evolution could design such a complex, ingenious and precise system. Even from a scientific standpoint, the simpler explanation is that our immune system, like everything else, was created by an all-wise, all-powerful Creator!

Prayer: I thank you, Father, for providing us with immune systems, and I ask you for good health. In Jesus' Name. Amen.

Ref: John Travis, "The Accidental Immune System," *Science News*, November 7, 1998, v.154, p.302

Hotheads Venting Heats Anger

Ephesians 4:31
"Let all bitterness, and wrath, and anger, and clamor, and evil speaking, be put away from you, with all malice."

For years, those filled with anger were advised to vent their anger. Supposedly venting one's anger—hitting a pillow, breaking things, yelling—would get rid of the anger. To many people, such advice seemed contrary to the spirit of Scripture. And a few more conservative voices in the counseling community maintained that far from dealing constructively with anger, venting actually increases it.

Now, a Duke University study has documented that venting not only increases anger, but is actually bad for one's health. In one study, researchers randomly assigned reading to 600 college students. Some students read pro-venting articles, others read anti-venting articles, while others read articles on unrelated subjects. Students were asked to write an essay on the article they read. Then, each was given negative comments about their essay—comments designed to make them angry—followed by the opportunity to hit a punching bag. Finally, each student was paired with an opponent in a competition that offered an opportunity for aggression. Researchers found that students who read a pro-venting article were twice as aggressive as the others. Aggression levels were also linked to how much the students liked hitting the punching bag. Other studies have shown that anger doubles or triples one's chance of having a heart attack, and long-term anger is linked to other health problems as well.

Science has again "learned" what the Bible has always taught: anger is not good for you or anyone else.

Prayer: Forgive me Father and take all anger from my heart so that I may show your love to me in Christ to those around me. Amen.

Ref: Judy Forman, "New advice for Hotheads: Cool It," *Star Tribune*, Sunday May 9, 1999, PE3

How Much Does Science Know?

Genesis 1:11
"Then God said, 'Let the earth bring forth grass, the herb yielding seed, and the fruit tree yielding fruit after his kind, whose seed is in itself, upon the earth'; and it was so."

What's left to be discovered? Many people have the idea that scientists have discovered just about everything when it comes to everyday subjects. Sure, scientists are searching for cures for many diseases and making discoveries of new subatomic particles. But when it comes to more common fields, say botany, even many scientists think everything has been discovered.

But a few botanists realize that even when it comes to wild plants in North America, a surprising percentage of plants still have not yet been discovered, named or described scientifically. One modern botanist who holds this belief has named 118 new plants, discovered 61 distinctive varieties and recorded 57 plants that have been named as new species. While some of the newly discovered plants that were found in remote locations, many are seen by people every day. A member of the lily family that was first described in 1972 grew only 10 miles from downtown San Francisco. What is now known as Morefield's leather flower was discovered in 1982 on a vacant lot in suburban Huntsville, Alabama. Botanists working in this field estimate that 5 percent of North American plants still remain either undescribed or undiscovered.

God created such a variety of plants that modern science still has not even discovered all the living plants around us. This truth should tell us that modern science cannot tell us that God did not create this wonderful variety and beauty.

Prayer: Father, we thank you for the variety and beauty that you have built into the plants around us. In Jesus' Name. Amen.

Ref: Susan Milius, "Unknown Plants under Our Noses," *Science News*, January 2, 1999, v.155, p.8

Evolutionary Prediction Fails

Genesis 1:24
"Then God said, 'Let the earth bring forth the living creature after his kind: cattle and creeping thing and beast of the earth, each according to his kind'; and it was so."

Scientists who believe in creation often point out that when we finally learn the genetics of a creature, that information doesn't match its supposed evolutionary history. There are many examples of this, and a new example was reported in the scientific literature at the end of 1998.

Evolutionary scientists had theorized that turtles were the last survivors of a very early group of creatures that later evolved into reptiles, birds and mammals. Their reasoning for this lies in the type of skull turtles have. Most reptiles and birds have two holes on each side of their skulls, behind their eyes. But turtles stand alone in having none of these holes. So they were classified as a separate evolutionary branch. But a comparison of the genetics of the turtle and other reptiles now places turtles among reptiles. They are not related to any supposed ancestor of reptiles, but rather are closest to alligators and birds. While some evolutionists are critical of this study, the study itself is the second that places turtles among the modern reptile family.

As more genetic information has become known in recent years, the relationships revealed have not generally been what evolutionists expected. And this is exactly what those who believe in the Creator would expect. So next time someone tries to tell you that belief in the Creator is ignorant and unscientific, just remember the turtle!

Prayer: Dear Father in heaven, I thank you that true knowledge glorifies you. Help my life and witness glorify you, as well. In Jesus' Name. Amen.

Ref: R. Monastersky, "Turtle genes upset reptilian family tree," *Science News*, December 5, 1998, V.154, p.358

Social Spiders

Job 8:13-14

"So are the paths of all who forget God; and the hypocrites hope shall perish, whose hope shall be cut off, and whose trust, shall be a spider's web."

A family reunion among spiders is not a pretty thing. Many of the some 35,000 species of spiders begin eating their siblings as soon as they hatch. For this reason, spiders are typically lone hunters. Evolutionists have always explained that spiders evolved this way because it works best for them.

But a few species of spiders are social. They live together in large communities, caring for each other's young and working together to get food. Their social structure is not like that of bees or ants; instead, spider society is more like a herd of wildebeests.

A South American social spider lives in multi-generational colonies that can have hundreds of thousands of members. Australia's crab spider is also social, building a nest of eucalyptus leaves. The colonial orb-weaving spider of Mexico and the U.S. Southwest builds huge colonies out of individual orb webs. The largest colony web ever found was 12 feet deep, six feet high and 600 feet long. It contained hundreds of thousands of members!

Evolutionary scientists have been struggling not just with an explanation for how social spiders evolved, but how it happened independently among eight unrelated families of spiders. But the answer should be obvious. The same Creator Who made the wildebeest a herd animal, also made social spiders the same way. He did it so we might recognize His existence, and then learn of His love for us in sending His Son be our Savior.

Prayer: We praise you, Father, for the amazing creativity with which you have signed your creation so we might seek you. Amen.

Ref: Laura Helmuth, "Spider Solidarity Forever," *Science News*, May 8, 1999, v.155, p.300

An Unused Religious Icon?

Psalm 9:17
"The wicked shall be turned into hell, and all the nations that forget God."

Are we Christians becoming less able to defend the truths of Scripture, like creation? Dozens of studies show that not only are we, as a group, less ready to defend the hope that we have in Christ, but also we are less able to explain why we have that hope.

George Barna, president of the Barna Research group has analyzed data from 28 religion surveys that he conducted between 1987 and 1996. In 1992 nearly 50 percent of those polled said that they had read the Bible during the previous week. By 1996, that figure had dropped to 34 percent. Yet, in another 1996 poll, 80 percent of those polled said the Bible is the most influential book in human history, and 76 percent said that the Bible is important to them. Barna concludes from this kind of information that for many Americans, the Bible has turned into an unused religious icon. The result is a general biblical ignorance. Barna documents this fact with examples from other surveys. For example, 80 percent of those asked thought, incorrectly, that the statement, "God helps those who help themselves" was in the Bible. In another instance, 50 percent of those polled thought that Jonah was a book of the Bible while 10 percent thought that Joan of Arc was Noah's wife.

If we are to defend the great truths of the Bible—like creation and the flood—and show their relevancy to Christ's saving work, we must each look at ourselves. Shouldn't we all study the Bible more than we do?

Prayer: Dear Father, increase in me a desire to study Your Word so that, with Your help, I can be more prepared to witness the hope of salvation I have in Jesus Christ. Amen.

Ref: David Briggs, "Survey: Bible mere icon in homes, *Brainerd Daily Dispatch* (Minn) Friday, August 2, 1996, p.9A

This Worm Gives God the Glory

Job 36:3
"I will fetch my knowledge from afar; and will ascribe righteousness to my Maker."

Scientists have finally completed the first genetic map of an animal, and the results have the signature of our Creator all over them. This newly mapped animal is the nematode, a very small worm. Even though it has only about 1,000 cells, it is very much like more complex animals. It has a nervous system, a brain, and it reproduces sexually.

Scientists had thought that the six-chromosome worm would have about 6,000 genes. But after they examined the 97 million bases in its DNA, they discovered that it has about 20,000 genes. They believe that about 3,000 of those genes are essential for all animal and human cells. This means that as scientists learn about these genes, they will also be gaining knowledge about those same genes in our bodies. The most important genes are clustered in the center of each chromosome, while the less essential genes are positioned at the end of each chromosome. They noted that the most important of the genes have more protection in the center of the chromosome.

Not only is the genetic code of this nematode much more complex than originally thought, but also it is elegantly designed. The genetic code itself is a sophisticated information system which could never have come about by chance. And now we learn that our all-wise Creator has also given special protection to the most essential of our genes by placing them in the middle of the chromosomes. The more we learn about how we are made, the more of the glory of our Creator we see in His excellent work.

Prayer: Father, I thank you that we are so wondrously made. Help me never fear that new knowledge will take your glory away. Amen.

Ref: J. Travis, "Worm Offers the First Animal Genome," *Science News,* December 12, 1998, v. 154, p. 372

One of Sin's Effects on Frogs

Romans 8:22
"For we know that the whole creation groaneth and travaileth in pain together until now."

Researchers have been trying to find out what is causing frogs to grow extra legs, or in some cases, grow no legs at all. While frogs with extra legs or no legs have been reported for centuries, there was some question about whether the incidence was increasing. Some suggested human pollution was to blame, while others suggested an increase in ultraviolet radiation was causing the problem. But none of the suspected pollutants have yet been found in association with deformations in the wild. Now, some answers are beginning to emerge.

Two different research papers point to a parasitic flatworm as the culprit. A study of deformed frogs in California showed that the ponds with deformed frogs also contained a species of snail that hosts the flatworm. A second study in the Pacific Northwest showed that the pattern of malformation in frogs there followed the pattern that is typical of parasite damage. Researchers there said that the pattern rules out suspected human pollutants called retinoids. Other researchers, who still suspect human pollution, point out that flatworms will probably not be the cause of frog deformation in every case.

We know that the real cause of these frog deformations is ultimately human pollution. That pollution is called sin. It was man's sin that brought decay and death to God's original perfect creation, because disobeying God does have consequences. The whole creation groans under the consequences of man's sin. That's why God sent His Son, Jesus Christ to deliver us from the ultimate consequences of our sin. Someday He will return to deliver all believers and the creation from sin forever.

Prayer: Father, help me be prepared for the return of my Lord Jesus Christ. Amen.

Ref: S. Milius, "Parasites make frogs grow extra legs, " *Science News*, May 1, 1999, v.155, p.277

Why Are You?

Genesis 1:27
"So God created man in his own image; in the image of God
created he him; male and female created he them."

When you know who made something and why it was made, you
know the purpose of the item. That's true of can openers, and it is true of
human life. Anyone who doesn't know who designed us or why; is
almost certain to be confused about his or her purpose in life.

For generations, evolutionary scientists tried to explain our
origin by comparing fossilized bones. But more recently they have
turned to comparing the genetic material of current human groups. You
probably remember a few years ago when these scientists announced
they had discovered that all living people could be genetically traced to a
woman in Africa many long evolutionary years ago. They called her
"Eve." Today they admit that this picture is a little simplistic.

Further studies have led to two contradictory theories. The first
theory asserts that all humans today evolved from a single African
population. The second position holds that modern humans are the result
of the interbreeding of populations from Africa, Asia and Europe. Some
also admit that the flaw in both positions is that the same genetic data can
be interpreted to support either theory.

The major flaw in all of these studies is that they assume man is
a product of evolution. As long as evolutionary science refuses to
consider God's account, which says He made us, it will be without a solid
scientific basis to answer questions about why. And people who do not
know why they were made are almost certain to live empty purposeless
lives.

Prayer: Dear Father, I thank you that you have made me and
that you have given me forgiveness through Jesus Christ.
Amen.

Ref: Bruce Bower, "DNA's Evolutionary Dilemma," *Science News*, February 6, 1999, v.155, p.88

A Walk on the Dark Side

Ecclesiastes 8:16-17b
"When I applied mine heart to know wisdom, and to see the business that is done upon the earth: (for also there is that neither day nor night seeth sleep with his eyes:) Then I beheld all the work of God, that a man cannot find out the work that is done under the sun:"

A scientific mistake has led to several discoveries about bumblebees. As scientists closed up their lab for the night at the State University of New York at Stony Brook, one of them accidentally left on an infrared TV monitoring system. When the researchers returned the next morning, they discovered that the bumblebees they had been studying were slipping out of the nest in the dark for a midnight snack. No, they didn't fly. Instead, they walked to the food supply the scientists had provided.

This discovery led the scientists to further study how bees navigate in the dark. When researchers moved their feeder, the bees walked through the dark to its old location. When scientists changed the position of the floor on which the bees walked, they followed their old path, even though it went in the wrong direction. After scientists cleaned the surface on which the bees walked, the bees still went in the compass direction of the original position of their feeder.

These results have led the scientists to believe that bumblebees navigate both by smell and by some sort of internal magnetic compass. Neither of these facts was known before. One final discovery is that bee feeding has two peak times, midday and midnight. Clearly, there is a lot more to learn.

God's creativity in providing all His creatures with the abilities they need appears to be unlimited. No matter how sophisticated science becomes, it will never run out of things to learn about God's creativity. This is another evidence that we are not the product of chance evolution.

Prayer: We thank you, Father, for your wonderful creativity in all things. Amen.

Ref: S.M. "Night life discovered for bumblebees," *Science News*, v.155, p.78

Two Languages that May Support Biblical History

Genesis 11:7
"Go to, let us go down and there confound their language, that they may not understand one another's speech."

The Bible's account of human history leads to the conclusion that the languages of the world had their birth at the Tower of Babel about 4,300 years ago. Evolutionists, however, place the separation of a Siberian and native North American group at thousands of years before that. Yet one linguist has found remarkable similarities between these two language groups that suggest these groups separated from one another much more recently.

The usual evolutionary model has native North Americans crossing a land bridge from Siberia to populate North America at least 10,000 years ago. So, one linguistic researcher wondered if there were any Siberian Languages that were related to North American languages. He compared the Ket language with the Na-Dene family of languages. The Ket language is spoken only in central Siberia and appears to have no linguistic relatives. The Na-Dene family of languages is spoken in western Canada and Alaska.

The researcher found that the two languages have 36 protowords that sound the same and have the same or similar meanings in both languages. Among the similarities are terms for body parts, natural phenomena, like snow and animals, and every day things. This evidence strongly suggests that the two languages are related; however, other linguists are not convinced.

If Ket and Na-Dene are truly related to each other, then surely these language groups separated more recently than the usual evolutionary model suggests. This conclusion would support a literal reading of the Biblical history.

> *Prayer: We thank you, Lord, that your Word is trustworthy in all that it says so that we can trust your promise of salvation. Amen.*

Ref: B. Bower, "Tongue ties across continents draw fire," *Science News*, November 14, 1998, v.154, p.311

Biblical "Kinds" Are Not Necessarily "Species"

Genesis 1:11
*"Then God said, 'Let the earth bring forth grass, the herb
yielding seed, and the fruit tree yielding fruit after its kind,
whose seed is in itself, upon the earth'; and it was so."*

Back in the 1970s, San Franciscans decided to restore their salt
marshes. Cordgrass was among the plants that were being restored to the
marshes. But, the cordgrass that was used in the restoration was a
commercially available species from the eastern United States, not the
California species. At the time, those working on the restoration did not
realize that the two almost identical species have some important
differences.

The eastern species, which is taller than the California species,
produces 21 times as much pollen as the California cordgrass. The
eastern species also produces eight times as many seeds as the California
species. But the two species are similar enough to cross-pollinate and
produce a fertile hybrid with many of the eastern species' characteristics.
Both the eastern species and the hybrid are able to colonize much closer
to the water than the native California species.

This battle between the eastern and California cordgrass
illustrates some scriptural truths. First, the "kinds" that Scripture says
everything is supposed to reproduce after are not the same as species.
"Species" are a human designation while "kinds" are God's designation.
These two species of cordgrass are obviously of the same kind since they
produce fertile hybrids. And since the eastern species is more robust than
the California species, it could well be closer to the original kind that
God created.

***Prayer: Father, I pray that I may truly have the characteristics
of your child through Jesus Christ, my Lord. Amen.***

Ref: S. Milius, "Superstud grass menaces San Francisco Bay," *Science News*, November 14, 1998,
v.154, p. 310

When Is a Billion Years Not a Billion Years?

Genesis 5:5
"So all the days that Adam lived were nine hundred and thirty years; and he died."

We often hear evolutionary scientists talking about this or that fossil that is, they say, "millions" or "billions" of years old. On the other hand, the Bible preserves ancient calendars from the time of Adam, such as in Genesis 5, which allow us to place the creation at only a few thousand years ago. How do evolutionists arrive at those millions of years?

All dating systems in use are based on the supposed evolutionary progression of life. Some people think that there are radioactive dating methods that are independent of the evolutionary tree, but the technical literature admits that this is not the case. Rock samples are dated by what are called index fossils. If a given rock sample has a certain shell in it, and the evolutionary tree says that creature lived 55 million years ago, the rock is dated at 55 million evolutionary years.

This dating system, however, doesn't always work the way expect it to. The rock sample, from a layer previously dated from other samples, was said to be 1.1 billion years old. Yet it had worm burrows in it. Evolutionist history says that worms didn't evolve until half a billion years ago. Another evolutionist examined the rock and said he found shells in it that date it at 545 million years ago. Yet other scientists looked at the rock and said that the shells weren't really shells. So they are now stuck with dates that conflict with each other by more that 100 percent.

Rocks will continue to be found which don't fit into the evolutionary timetable because they, and all life, were formed by God not millions but only thousands of years ago.

Prayer: I thank you, Father, that you have made everything and told us about it in your perfect Word. In Jesus' Name. Amen.

Ref: R.M. "A rock that lies about its age," Science News, November 21, 1998, v. 154, p.332

Two Pairs of Ears Offer a Telling Story

Psalm 78:1
"Give ear, O my people, to my law; incline your ears to the words of my mouth."

Did you ever wonder what those folds in your ear do? Scientists have recently confirmed their suspicions that those little folds, called pinnae, help you tell where sounds are coming from. All those little folds amplify or weaken certain frequencies, and your brain processes these changes and figures out where the sound you hear is coming from.

Now comes the mystery. Your pinnae are unique to you. Pinnae have been compared to fingerprints—each person has unique pinnea. This means that your pinnae do different things to the sounds you hear than, say, what your wife hears. Still, if we both hear the same sound, we can both tell where it is coming from.

Scientists have theorized that the brain is designed to learn the unique shape of its pinnae, and from that, to learn how to identify the source of a sound. This theory was recently confirmed when scientists gave volunteers plastic molds designed to change the shape of their pinnae. After the molds were implanted into their ears, the subjects discovered that it took several weeks for them to develop the ability to discern the height of a source of sound. Even more interesting is that when the molds were removed, the subjects did not have to relearn how to tell the height of a sound. One researcher compared this to learning a new language. The brains were now able to effectively use two different sets of pinnae.

This ability to relearn how to tell the height and direction of sound, even if our pinnae are injured, points to a wise Creator Who provides for our needs. Such a contingency plan as this could not have come from mindless and loveless evolution.

Prayer: I thank you, dear father, for the ability to hear and read your Word so that I can learn more about my Savior. Amen.

Ref: J.T. "The Brain gets a (new) earful," *Science News*, November 14, 1999, v.154, p. 316

Has Aaron's Y Chromosome Been Found?

Exodus 40:13
"And thou shall put upon Aaron holy garments, and anoint him and sanctify him, that he may minister to me in the priest's office."

According to biblical chronology, God ordered Aaron to be consecrated as the first priest of Israel about 3,400 years ago. From Aaron, priesthood was passed down from father to son until the official priesthood ended with the destruction of Jerusalem in 70 AD. After that, fathers would tell their children that they were descended from the priestly class. Though only these oral histories exist today, modern Judaism still honors Aaron's descendants. Today they are called *cohanim*, which is the plural of *cohen*, meaning, "priest" in Hebrew.

A modern geneticist wondered if there was any way to prove the oral history. He reasoned that if all *cohanim* today are descended from Aaron, and if the oral history is true, modern *cohanim* should share some distinctive sections of DNA on their Y chromosome. (The Y-chromosome is always passed down from father to son virtually unchanged.) The geneticists first study involved two genetic markers of 188 Jewish men from Israel, Canada and the United Kingdom. Of this number, 68 said they were *cohanim*. The results of this first study showed that the *cohanim* share Y chromosome material that is quite distinct from other Jews. A second study involving 12 markers on the Y chromosome of 306 Jewish men gave similar results—the *cohanim* did have a distinctive Y chromosome. This chromosome evidence strongly indicated that both the biblical and modern oral history of the Jewish priesthood is completely trustworthy.

Prayer: Dear Father, I thank you that I have been made your priest in Christ. Amen.

Ref: John Travis, "The Priests' Chromosome?" *Science News*, October 3, 1998, v.154, p.218

Earthquake Caused by Noah's Flood?

Genesis 7:11
"In the six hundredth year of Noah's life, in the second month, the seventeenth day of the month, the same day there all the fountains of the great deep were broken up, and the windows of heaven were opened."

Miles deep within the Earth, heat and pressure force water out of certain minerals, changing or metamorphing the rock. All the textbooks say that it takes millions of years for such metamorphic rock to form. New research shows pretty convincingly that all the textbooks are wrong.

Normally, earthquakes originate in the top 12 kilometers of the earth's crust. Deeper quakes, especially where there are no known faults, have been mysteries. The At 18 kilometers deep the 1989 Loma Prieta earthquake was one such mystery. But a new discovery by three Yale University geologists may explain these deeper quakes and support the Bible's historical chronology.

The standard model for metamorphic rock formation says that heat and pressure deep underground slowly force water from the crystals that form the rock. This process takes place so deeply in the earth that the rock itself flows plastically. Now, the Yale geologists studying the process have discovered that under these conditions tiny droplets of water are expelled from the crystals much more rapidly than previously thought. In fact, the water is expelled so rapidly that it builds enough pressure to actually fracture the rock.

The geologists who presented this research say that what was thought to have taken millions of years can happen in mere centuries or even decades. And this would easily allow for the origin of some of today's metamorphic rock to be traced back to the Flood of Noah on the time scale provided by the Bible.

Prayer: Father, I thank you for your trustworthy Word in this sin-filled world. Amen.

Ref: S. Simpson, "High-pressure water triggers tremors," *Science News*, November 21, 1998, v.154, p.327

Freak of Nature or Divine Design?

Genesis 1:14
"Then God said, 'Let there be lights in the firmament of the heavens to divide the day from the night; and let them be for signs and seasons, and for days and years..."'

We recently reported on this program that another set of new planets had been discovered circling another star. We pointed out that in each case, most of the planets were too large to support life. In many of the instances, the planets also had egg-shaped orbits that would provide them with harsh, highly erratic climates—if they had any atmosphere at all.

In his recent announcement of the discovery of the 17th planet discovered orbiting another star, Geoffrey W. Marcy of San Francisco State University and the University of California, Berkeley, points out another problem with these solar systems. In pointing out the problem, he wonders if our solar system, where the planets have relatively circular orbits, is an exception to the rule.

In every other planetary system so far seen, the planets either have very oval orbits or they lie closer to their star than Mercury is to our sun. Marcy points out than an additional problem with having a planet with a highly elongated orbit in a solar system is that it would eliminate any planets with circular orbits. It would be only a matter of time before the planet with the elongated orbit would send the planet with a circular orbit into its star or out into deep space.

But where Marcy is wrong is that while our solar system may be unique it is no freak of nature. Rather, our solar system has clearly been specially designed by a loving and all-powerful God to support life.

Prayer: I thank you, Father that you have created the solar system as a good place to live and learn of your love for me through Jesus Christ. In His Name. Amen.

Ref: R.C., "Solar system planets: Freaks of nature?" *Science News*, January 30, 1999, v.155, p.79

Real Ant Farms

Proverbs 6:6-8
"Go to the ant, thou sluggard; consider her ways and be wise:
Which, having no guide, overseer or ruler, provideth her meat in
the summer, and gathereth her food in the harvest."

There are more than 200 ant species in the attine family of ants.
All of them farm for a living. And they are as sophisticated about their
farming as any modern human farmer. They are found anywhere from
the tropics to New Jersey. The specialization shown by these ants makes
an evolutionary explanation for them impossible.

One example is the famous leaf-cutting ants found in the tropics.
Leaf cutters use the leaf pieces they collect to make mulch. Once the
mulch is ready, they grow a fungus in it, which they eat. That fungus
grew from a "seed" fungus that was stored in a special pouch in the
mouth of the queen who began that colony and she brought it from the
colony where she grew up. That she has this special pouch for seed
material for her children is a witness to God's intelligent design.

The ants' digestive system is specialized, too, so that its waste is
a ready-to-use fertilizer for the crop.

Moreover, the ants weed their crop. Certain ants inspect the
fungus ball searching for invading molds. When one is found, the ant
will remove it. To make sure that she does not return any contaminating
bit of the mold to her crop, the ant then cleans herself before returning to
the crop. This, again, is a powerful witness to divine design. The ants
also generate herbicides and antibiotics which they use as needed on the
crop, another witness to design. Finally, they also prune their crop in a
way that has been shown to improve its productivity.

All these special design features and behaviors seem to have
been put into one creature as a witness against evolution.

Prayer: I praise and glorify you, Father for the evidence of
your great glory. Amen.

Ref: Susan Milius, "Old MacDonald Was an Ant," *Science News*, November 21, 1998, v. 154, p. 334

43

Researchers Find a Hidden Cost to the Internet

Proverbs 18:24
"A man that hath friends must shew himself friendly, and there is a friend who sticketh closer than a brother."

Have you been on the Internet yet? If so, it may be costing you more than you think. That's the suggestion of a study done by researchers from Carnegie Mellon University in Pittsburgh.

The study involved 169 people in 73 households who did not have a connection to the Internet. They completed questionnaires and were interviewed in their homes at the beginning of the two-year study, during the study and at the end. They were given computers and software, e-mail accounts and access to the Internet at no cost to them. In return, the test subjects allowed researchers to monitor family members' Internet use.

At the end of the study, researchers found that those who used the Internet and e-mail frequently showed small but important decreases in the time spent with family members. They also had a smaller circle of friends. Frequent Internet and e-mail users also showed increased loneliness and signs of mild depression. The greater the computer use, the greater the loneliness and depression.

These results were not affected by race, age, sex, or income; however, this effect was greatest for teens. The researchers suggested that when teens feel isolated and lonely they are more likely to escape to the Internet.

Perhaps the best suggestion in light of these findings is to limit Internet time, just as television time is limited for many young people. The extra time could be used as the book of Proverbs advises: "A man who has friends must himself be friendly..." In other words, it is healthy to cultivate friendships.

Prayer: I thank you, my Lord Jesus, for my friends. And I thank you that you have been my best friend in saving me. Amen.

Ref: B.B., "Social disconnections on-line," *Science News*, September 12, 1998, v. 154, p.168

Microbe Argues for God's Unlimited Creativity

Genesis 1:20
*"And God said, Let the waters bring forth abundantly the
moving creature that hath life, and fowl that may fly above the
earth in the open firmament of heaven"*

Aluminum, the most abundant metal in the Earth's crust, is
usually thought to be poisonous to life. When aluminum becomes too
concentrated in soil, it will greatly reduce crop yield. No known living
creature uses aluminum in its metabolism. That's why many scientists are
skeptical of claims by a researcher from the University of Maryland
Biotechnology Institute in Baltimore. She believes she has discovered a
microbe that not only tolerates aluminum, but also needs it to live.

The scientist had been hiking near hot springs in Yellowstone
National Park when she noticed what looked like aluminum silicate, (and
later proved to be just that). It was being carried by the water over a dark
green microbial mat on the rocks. She collected a sample of the mat, and
she and her fellow scientists later found a microbe whose growth seems
to require aluminum. It grows best in water that's about 160 degrees
Fahrenheit. While researchers don't know what the microbes use the
aluminum for, they have found that the higher the temperature of the
water, the more aluminum it seems to need. They think that the
aluminum may be used as the organism's source of energy.

If this organism proves to use aluminum, it would be an example
of God's creativity. With His unlimited creativity all around us, one
doesn't need to go to science fiction to find creatures that work in
seemingly impossible ways and ways that could never have come about
by chance evolution.

***Prayer: In wonder and awe, I praise you for your creativity,
dear Father. Amen.***

Ref: J. Travis, "Novel bacteria have a taste for aluminum," *Science News*, May 30, 1998, v. 153, p.
341

"Wo Fe Fe"

Acts 2:6
"Now when this was noised abroad, the multitude came together, and were confounded, because that every man heard them speak in his own language."

New research suggests that you might want to be careful about what you say around babies. As every parent knows, young children are sponges that easily absorb knowledge from their surroundings. Now, there is research that indicates that by seven months of age, infants have already absorbed the basic rules of language.

In this study, seven-month-old infants listened to a sequence of syllables in which the third syllable was the same as the first. One that was used was "ga to ga." After listening to different versions of the same pattern, the infants were given a new sequence in which the last two syllables were the same, such as "wo fe fe." Researchers found that babies looked in the direction producing new sequences much longer, showing curiosity and surprise at them. Further study of the infants' responses showed that they were responding to new sounds and not just changes in the pattern of syllables. This study supports the idea that we are born with brain circuits that are, pre-wired to learn and use grammar, say researchers.

The question for those who believe that we are the product of chance evolution is: How did we get pre-wired for learning and using grammar if we are the result of chance forces? The answer is, of course, that when God created us, He not only gave us the gift of language, but He gave us brains that are eager to learn language from infancy. Then he gave us His Word of salvation in Jesus Christ, so that He would not lose us.

Prayer: I thank you for the gift of language, dear Father, and the ability to learn it so that I could learn of my Savior. Amen.

Ref: B.B. "Follow the rules, baby," *Science News*, January 16, 1999, v. 155, p. 42

Evolution Can't Digest This Fly

Psalm 148:10,13
"Beasts and all cattle; Creeping things and flying fowl. Let them praise the name of the LORD: for his name alone is excellent; his glory is above the earth and heaven."

Conifers aren't bothered by too many insect pests. That's because the oils that give them that nice pine scent are poisonous. They would be poisonous to the conifer sawfly larva if it digested them, too. But while it munches the pine needles, its body is busy storing the poisonous oils for future defense. The fact that the sawfly not only eats pine needles, but also uses the poisonous oils for defense poses a difficult problem for evolution which appears to have no solution.

The sawfly larva's unique ability makes it a serious conifer pest both in North America and Europe. Part of the larva's secret is its digestive system. It separates the poisonous oils in the pine needles from the nutritious pulp. Then it regurgitates the poisonous oils and stores them in two special sacs in its mouth. These sacs are lined with a chitinous material that protects the rest of the mouth from the acidic, poisonous oils. When threatened by a spider or even a bird, the sawfly releases some of this bad smelling oil in the direction of the threat. This strategy effectively convinces most aggressors to seek a meal somewhere else.

The problem for evolution is that these complicated structures and special abilities could not have developed in a gradual, step-by-step fashion as evolution demands. If the sawfly once did not originally eat pine needles, then how did it acquire the special ability to eat them and develop this special defense system? So, even this humble fly bears witness to its Creator.

Prayer: Dear Lord, help me to be a better witness of your love for me. Amen.

Ref: **Creation Research Society**, May/June 1999, "Neodiprion (Sawfly)" Mark Stewart

1,300 Year-Old Recording Discovered?

Genesis 4:21
"And his brother's name was Jubal: he was the father of all such as handle the harp and organ."

If you've ever seen any photographs of Mayan pyramids, you will probably remember that they typically have a large set of stairs going to the top. Now an acoustics expert offers some interesting evidence that these steps may in fact be the world's oldest recording.

If you stand before the stairway of the Pyramid of Kukulkan at Chichen Itza, Mexico and clap your hands, you will hear a curious, descending echo. If you are familiar with the quetzal, whose bright green and red feathers the ancient Mayans treasured, the echo will remind you of the call of this bird.

When sonograms of the quetzal's call were compared to sonograms of the echo returning off the steps of this pyramid, they were surprisingly close. Both begin at a frequency of about 1,500 hertz and fall at the same rate to less than 1,000 hertz! Making this even more interesting is the fact that the pyramid itself has a picture of Quetzalcoatl wearing a coat of quetzal feathers. According to legend, Quetzalcoatl was half quetzal. Traditional Mayan scholars don't doubt that the Mayans were clever enough to have purposely built the pyramid to provide this echo, and acoustics experts have found a similar echo at a temple in Uxmal, Mexico.

The Bible tells us that by the eighth generation of human beings musical instruments were in use. This pyramid may be evidence that we were created with the aptitude to understand and use sound, just as the Bible depicts in its opening chapters.

Prayer: I thank you for the gift of sound and music, dear Father. Let me always be filled with praises for you. Amen.

Ref: Peter Weiss, "Singing Stairs," *Science News*, January 16, 1999, v.155, p.44

What Did Dinosaur Skin Look Like?

Job 41:1, 30
"Canst thou draw out leviathan with an hook? or his tongue with a cord which thou lettest down? Sharp stones are under him: he spreadeth sharp pointed things upon the mire."

Before the Great Flood of Noah, what did our forefathers see when confronted with a dinosaur? What did dinosaur skin look like? What were their babies like? Biblical history places the creation of the great land-living dinosaurs on the same day as man was created. That means that there is little doubt that at least some of the people we read about in the Bible saw dinosaurs. Many modern creationists think that Job's description of leviathan actually describes some sort of dinosaur.

For generations, what we knew or thought we knew about dinosaurs was a result of studying their bones. But now, recent discoveries are offering some surprises about what dinosaurs actually looked like. One discovery is the first-ever-fossilized impression of the skin of a duckbilled dinosaur known as a hadrosaur. This interesting fossil, found in southwest New Mexico, shows that the hadrosaur had bumps on its skin that averaged a little over half-an-inch in diameter. Each rose next to each other like little mountains. This means, as one scientist put it, that if you were to pet a hadrosaur, it would feel like running your hand over a mountain bike tire.

In Patagonia, scientists have discovered the first known embryos, complete with skin impressions, of a sauropod, which was a large, four-footed dinosaur. These discoveries are giving us a more accurate picture of what these living monuments to God's power and creativity looked like. And one day we can ask Job exactly what leviathan was.

Prayer: Father, I thank you because your power and Godhead are evident in the creation around us. In Jesus' Name. Amen.

Ref: R. Monastersky, "Getting under a dinosaur's skin," *Science News*, V.155, p.38, January 16, 1999

Is Your Brain Shrinking Because of Stress?

Matthew 6:25
"Therefore I say unto you, take no thought about your life, what ye shall eat or what ye shall drink; nor yet for your body, what ye shall put on. Is not life more than meat the body more than raiment?"

If stress is an everyday part of your life, or if some past trauma has continued to haunt your thoughts every day, you may be losing part of your brain. Several studies have linked such constant or obsessive stress with a shrinking of the part of your brain called the hippocampus. Studies show that under such conditions this part of the brain can become 25 percent smaller than normal. These findings should get our attention, since we use the hippocampus for long-term memory and conscious memory. (The hippocampus is the same part of the brain attacked by Alzheimer's disease.)

The good news is that once we remove the stress, or deal constructively with past trauma, the hippocampus returns, with time, to normal size. Several theories have been offered by researchers to explain why the hippocampus shrinks under stress; however, other researchers are not convinced that stress causes the shrinkage.

Is it possible for a Christian to avoid the damage that stress produces? Look at all that St. Paul endured for the sake of the Gospel, and yet all of his writings show no evidence of unusual memory loss, which is common when one is under constant stress. One way is to avoid stress damage is remove the cause of the stress. If that's not possible, another strategy is to learn how to deal with past traumatic events. And what better way to do this than to remind ourselves of the Lord's past faithfulness to us, and, in prayer, commend all our fears and stress into His hands. Science is learning that life is indeed more than food and clothing.

Prayer: I cast all my cares, fears and worries on you, oh Lord. Amen.

Ref: Robert Sapolasky, "Stress and your shrinking brain," D*iscover*, March 1999

Not Such Bumblebees

Psalm 98:4
"Make a joyfully noise unto the LORD, all the earth; make a loud noise and rejoice, and sing praise."

Scientists have long known that honeybees communicate the location of nectar sources to their nest mates. The language that many species use has even been deciphered. At the same time, it was long accepted that bumblebees did not communicate the location of food resources to their nest mates. After all, it is astonishing enough that a supposedly simple creature like the honeybee possesses the advanced skill of communication, but can even simpler bumblebees communicate?

Now, research reveals that there are at least several species of bumblebees and other stingless bees that communicate among themselves. As might be expected, the languages differ from species to species. One species of bumblebee reports her find to her nest mates by running zigzag over the combs and bumping into her nest mates. This leads them to check out the quality of the nectar she just returned. If she has brought several good loads in, they join her in bringing in more. Other species lay scent trails to the food source. When a member of another species finds a good source, she communicates to her nest mates through a series of pulsing buzzes. As she communicates the height of the source by the speed of the pulses, she performs a circular dance that communicates the rest of the instructions.

The Bible is accurate when it says that the entire creation can offer God praise. The languages may differ; the form of communication may not even look like communication to us. But now we know that since communication is a gift from God, even so-called simple creatures can be given the gift by their Creator.

Prayer: Father, I too praise and thank you for the gift of communication. Amen.

Ref: Susan Milus, "Look Who's Dancing," *Science News*, v.155, p.216, April 3, 1999

There's Not Chance It's Chance!

Psalm 104:24
"O LORD, how manifold are thy works! in wisdom thou hast made them all: the earth is full of thy riches. . . ."

As we learn more about the genetic code, it is becoming increasingly difficult to escape the conclusion that all living things—and by implication, all things—have a very wise, personal Creator. At least one researcher in this field has admitted that it is extremely unlikely that the genetic code shared by almost all living things arose by chance.

Researchers have concluded that nearly a billion billion genetic codes are possible. That's a one, followed by 20 zeros! But not all genetic codes are created equal. Some are better than others at preventing errors when new genetic material or protein made by genetic material is produced. And when genetic errors are made in the production of protein, a better genetic code will seek to minimize the error. The genetic code we have is the best at doing this. More than that, the code we have is the best at minimizing the damage caused by faulty proteins at the very genes where it is most likely to happen! After admitting this code couldn't happen by chance, one researcher then gave the credit to "natural selection."

That the genetic code is an information storage system vastly more complex anything we have ever built should tell us that it has a wise and powerful Creator. That the genetic system we have is the best one of over one billion billion possibilities should seal the case for God as Creator.

Prayer: I thank You, Father, that I am your hand-made possession. Amen.

Ref: Jonathan Knight, "Top Translator," *New Scientist*, 18 April 1998, p.15

God Extends the Olive Leaf to Us

Genesis 8:11

"And the dove came in to him in the evening, and, lo, in her mouth was an olive leaf pluckt off: so Noah knew that the waters were abated from off the earth."

Many plant species undoubtedly became extinct during the Great Flood at the time of Noah. It has been speculated that many of the plants lost forever during the Flood offered better nutrition or medications than many of the plants we know today. This could be why, after the Flood, God allowed the eating of meat for the first time. Meat could make up for the nutrition no longer available in the now extinct plants.

Today, we are extracting some powerful medicines from plants from all over the world. The olive is one plant that may hint at some of the extraordinary abilities that God may have built into some of the extinct plants. We've all heard about the healthy effects of olive oil. But it turns out that the olive leaf, too, provides some very healthy extracts. One extract, oleuropein, found in all parts of the olive plant, has been shown to delay the formation of poisons formed by mold growing in nut meal. It can also stop the growth of one type of staphylococcus. Other studies suggest that this extract may relax the smooth muscles of artery walls, temporarily lowering blood pressure. In addition, it may help the heart function better and inhibit the formation of clogging plaques in the blood stream. Other extracts from the olive leaf have been shown not only to be antioxidants, but also to slow the production of poisons or even destroy certain bacteria.

> **Prayer: I thank you, dear Father for all the blessings of the earth. Amen.**

Ref: *Optimal Nutrients*, Olive Leaf Extract, Foster, CA

An Evolutionary Quandary

Psalm 14:1

"The fool has said in his heart, 'There is no God.' They are corrupt, they have done abominable works, there is none who doeth good."

How did life get started? We all know that the Bible tells us that life got started when God created it. But we have also learned, perhaps in school, or reading the news, that evolutionists believe that life could have started in the distant past all by itself if the conditions were right. Perhaps you have read that scientists can form amino acids in the laboratory. Besides the fact that amino acids are a long way from life, the amino acids made in the laboratory are the wrong kind for life. The idea that life can form spontaneously under the right conditions places modern evolutionary scientists in a very precarious position.

One assumption of evolutionary scientists is that the early earth had no oxygen. Oxygen destroys amino acids. Yet geologists tell us that oxidation even in the oldest rocks proves that the Earth has always had an oxygen atmosphere. Besides, if the earth didn't have an oxygen atmosphere when the first amino acids formed, that means there was no ozone, a form of oxygen, to protect those amino acids. Amino acids are quickly destroyed by ultraviolet light from the sun. So, you see, with or without oxygen, amino acids can't begin to form. The truth is that the Bible offers the only sensible explanation for the origin of life. As the great scientist, Louis Pasteur said, "Life can only come from life." And the Bible tells us that eternal life can only come from the eternal life-giver, Jesus Christ.

Prayer: Father, in your wisdom, you have put to shame those who would deny that you. Thank you for salvation in Jesus. Amen.

Ref: David Rosevear, "The Myth of Chemical Evolution," **ICR**, *Impact #313,* July 1999

What Do Chickens Talk About?

1 Corinthians 15:49
"And as we have borne the image of the earthy, we shall also bear the image of the heavenly."

We have all heard that our ability to communicate our thoughts is one of the things that make humans unique. On the evolutionary scale of things, language was considered a sophisticated development. Supposedly, chickens didn't have any thoughts worth communicating anyway. Now, this view is being challenged by multiple studies of animal communication.

Most of the studies have looked to see if there is any pattern to various species' distress calls. For example, researchers have found vervet monkeys are very specific as they warn each other of danger. When the danger is a large monkey-eating creature like a tiger, the danger call is "wrr." At this call all the monkeys in earshot hurry up their trees. If an eagle threatens them, the alarm call is a grunt, causing the monkeys to dive into the underbrush. If the threat is a snake, the alarm call becomes chuttering sounds. This leads the monkeys to stand up and look around the ground for the threat. Researchers were amazed that specific sounds lead to specific actions, depending on the threat.

Other animals, like ring-tailed lemurs, have been found to have their own pattern of communication. And yes, even the lowly chicken has a different warning call, depending on whether the threat is a raccoon on the ground or a hawk in the air!

So it's not our ability to communicate that makes us unique, but the fact that we were created in God's image and we have been redeemed by Christ's blood.

Prayer: I thank you, dear Father, that you have made me to have a relationship with you, and that You have redeemed me. Amen.

Ref: Susan Milus, "The Science of EEEEEK!," *Science News*, September 12, 1998

Sunfish Have Personality, Too

Revelation 4:11
"You are worthy, O Lord, to receive glory and honor and power: for thou hast created all things, and by thy pleasure thy are and were created."

Are you shy or bold? Actually, if we are honest with ourselves, we must admit that we all have areas where we are bold, and other things we're shy about. A person can be bold in willing to fly in any aircraft, and at the same time terrified when crossing a high bridge.

Now scientists have surprised themselves by discovering that the same is true of sunfish, crabs and the octopus. Evolutionary scientists had always assumed that since "survival of the fittest" is the driving force behind evolution, evolution would favor boldness. But scientists found that when they introduced a novel object —a stick with a red tip— into a school of sunfish, about a quarter of the fish fled from it, while another quarter of the school boldly nipped at the stick. But those fish who were shy toward the stick were no more likely than any other individuals in the school to be shy when introduced to unfamiliar food. Those who nipped at the stick didn't prove any more likely to be bold when introduced to unfamiliar food. Other researchers have concluded that studies show the same behavior in crabs and in the octopus. The researchers marveled that even so-called simple animals showed individually unique traits just as humans do. As one said, "There must be something very special here."

There *is* something very special here. That these creatures have individual personalities says nothing about evolution. But it does remind us that all living things have but one divine Creator.

Prayer: Father, I thank you for my unique personality and ask that it would be better conformed to Christ, my Savior. Amen.

Ref: S. Milius, "Fish Nature: Sometimes shy, sometimes bold," *Science News,* v. 154, p.263, October 24, 1998

More Proof You're Not Accidental Chemistry

Psalm 139:14
"I will praise thee; for I am fearfully and wonderfully made;
marvelous are thy works; and that my soul knoweth right well."

Most people know that the most important molecule in our bodies is DNA. But another large molecule called ATP is what keeps you and your DNA alive. ATP is the primary energy system within your body. It generates the electricity in your nerves, making it possible to move your muscles. ATP's operation should remove any doubt that it is a product of a wise and intelligent Creator.

ATP converts the energy in the food you eat to just the right level needed most of the time by your cells. When ATP delivers its energy within your cell, it loses a group of phosphate atoms and becomes ADP. ADP is immediately recycled back into ATP within your mitochondria to again deliver its energy. On the average, each ATP is recycled like this three times per second. Despite the fact that each of your one hundred trillion cells have one billion ATP molecules, you would die within a few minutes if anything interfered with this recycling. During the course of a day, your body will recycle 400 pounds of ATP! Within your mitochondria are what have been described as molecular "water wheels." When working at its peak this "water wheel" spins as fast as 200 revolutions per second, generating 600 ATP molecules during that second!

The Psalmist was right when he praised God because he recognized that he was fearfully and wonderfully made.

Prayer: Lord, I thank you that I am fearfully and wonderfully made. Amen.

Ref: Jerry Bergman," ATP: The Perfect Energy Currency for the Cell," *Creation Research Society Quarterly,* v.36, p.2, June 1999

Too Much on Your Mind?

Proverbs 12:25
"Heaviness in the heart of man maketh it stoop, but a good word maketh it glad."

The Bible frequently warns against anxiety. Modern science agrees that anxiety produces all sorts of negative changes in the body. A recent study on rats suggests that the damage done by anxiety begins practically at the onset of anxiety.

In the study, scientists used electrodes on some of the rats to simulate what happens in the brain when learning takes place. Then they put all the animals through a learning exercise. The rats were to learn to find and swim to a submerged platform in a water tank. Scientists found that those rats whose learning centers were over stimulated did much more poorly at learning than those who had not been over stimulated.

Learning normally takes place as new experiences strengthen connections between nerve cells. In this study, scientists concluded that over stimulation prevents those connections from being strengthened. In practical terms, this means that there is a limit to how quickly we can absorb new information. In other words, the student who stays up all night cramming for final exams will end up less able to learn than the non-anxious student who sleeps peacefully the night before finals.

So, if you have ever had times when you thought that your brain just couldn't absorb more information, you were probably right. So now we have yet another reason why God, in His Word, urges us not to be anxious, but to cast all our cares on Him. After all, if He sent His only Son to rescue you from sin, death and the devil, will He not also carry you through whatever makes you anxious?

Prayer: Dear Lord, I cast all my cares on you, for you love me. Amen.

Ref: J.T., "Rats have too much on their minds," *Science News*, v.154, p.250, October 17, 1998

The Bible: Convincing History

Joshua 11:11

"And they smote all the souls that were therein the edge of the sword, utterly destroying them; there was not any left to breathe: and he burnt Hazor with fire."

The Bible tells us that in conquering Canaan, Joshua destroyed the city of Hazor, which was the chief city in the area. Bible critics are usually skeptical of the claims made by the Bible, especially when it comes to Israel's conquest of Canaan. Excavations at Hazor began in 1955 and have gone on sporadically ever since. Archaeologists wanted to know when Hazor was destroyed and who destroyed it? Doubting that Israel could have caused the destruction, some suggested Hazor was destroyed by the Philistines, by another Canaanite city, or by the Egyptians.

New excavations during the 1990s revealed some interesting answers. The city was clearly destroyed by fire, as attested by the remaining ashes of the city, which are 3 feet deep in some places. This is consistent with Scripture's account. Because of the large amount of olive oil stored in large jars in the palace, the fire was especially bad there, reaching temperatures estimated at over 2000 degrees Fahrenheit. In addition, statues and idols were smashed, which is what the Israelites typically did in obedience to God. The idols destroyed were gods worshiped by the Philistines, Canaanites and Egyptians, making them unlikely as the destroyers. The pattern of destruction is the same as described by Scripture, leaving the only conclusion, say those closest to the excavations, that only Israel could have done this! It's good to see that even those who doubt the truth of Scripture can be convinced of its truth by history itself.

Prayer: I thank You, Lord, that I can be sure of the Bible's promise of salvation. Amen.

Ref: Amnon Ben-Tor and Maria Teresa Rubiato, "Did the Israelites Destroy the Canaanite City?" *Biblical Archaeology Review*, May/June 1999, p.22

Deceit Is Cuckoo

Proverbs 20:17
"Bread of deceit is sweet to a man, but afterward his mouth shall be filled with gravel."

You may be aware that the common cuckoo does not feed or raise its own young. Instead, it lays its eggs in the nests of other birds. The adoptive parents feed and raise the young cuckoo as their own until the cuckoo gets larger than the foster parents and flies away without so much as a "thank you."

One common adoptive parent for the young cuckoos is the reed warbler, whose behavior pattern is quite different from the cuckoos. For example, reed warbler parents recognize hungry baby birds by their persistent calling. Cuckoos typically lay but one egg in an adoptive nest. Once this egg hatches, the young cuckoo throws the reed warbler eggs out of the nest. So how does one little baby cuckoo manage to convince the parent reed warbler that it is half a dozen reed warbler babies to be fed? Researchers have finally learned the amazing answer to that question. They say that the baby cuckoo fools its adoptive parents by sounding like as many as eight baby reed warblers. The act is so convincing that it gets all the food it wants.

Who teaches the baby cuckoos this trick? Certainly not the mother cuckoo who, incidentally, misses out on all the fulfillment of family life. The cuckoo reminds us that deceit robs us of good experiences in our lives. That's why it is comforting when our perfect God of truth tells us that He never changes.

Prayer: Forgive me, dear Father, for any deceit in my life, and help me to live a life of truth and honesty. In Jesus' Name. Amen.

Ref: L.H., "Cuckoos beg doggedly to trick hosts," *Science News*, v.155, p.158, March 6, 1999

Is Your Memory This Good?

Luke 12:24
"Consider the ravens, for they neither sow nor reap, which have neither storehouse nor barn; and God feedeth them: how much more are ye better than the fowls?"

Do you remember how much milk is in your refrigerator? Do you remember which leftovers you have in there, and where you put them? This kind of memory is called episodic memory. It allows you to travel back in time with your mind to remember the details of a past action. Until now, it was thought that only humans had this kind of memory, although some researchers were convinced that monkeys and rats had episodic memory.

Now, to researchers' surprise, several species of birds that store food have convincingly displayed episodic memory and more. The scrub jay's favorite food is waxmoth larvae. Researchers allowed some of the jays to learn that the larvae rot after a few days. Researchers gave scrub jays the larvae to hide in sand-filled ice cube trays. After five days the jays were given another favorite food–peanuts–to hide.

Later, the birds were allowed to collect their buried treasures. The birds who had learned that the larvae rotted after a couple of days didn't even bother to collect them. They looked only for peanuts, and remembered where to look for them. The jays who didn't know that the larvae rotted looked for them first.

In another research project, Clark's nutcrackers remembered where they had buried food morsels nine months earlier. This is something evolutionary scientists never expected in what they consider lower forms of life than we are. God has seen to the food needs of these species by not only providing them with food, but also by giving them extraordinary episodic memory like ours.

Prayer: I remember and thank you for all your goodness to me, Lord. Amen.

Ref: S. Milius, "Birds can remember what, where, and when," *Science News*, v.154, p.181

Jewel of Death

Genesis 2:17
"But of the tree of the knowledge of good and evil thou shalt not eat of it: for in the day that thou eatest thereof thou shalt surely die."

The consequences of man's sin have touched every part of the creation. After God made us, He declared His entire creation "very good," which by God's standards is perfect. But human rebellion against God set at work the forces of death throughout the creation. The gnat-like jewel wasp is a good example of this corrupted creation. Under the microscope the jewel wasp lives up to its name, showing patterns of iridescent colors that change with the light. But that's about all there is about the jewel wasp that reflects its perfect creation.

The female jewel wasp, ready to lay her eggs, seeks out fly pupae, most readily found in the corpses of dead animals. There she injects the fly pupae with venom to kill them and then lays 20 to 40 eggs in each pupa. After one or two days, the eggs hatch and the hatchlings feed. Then males, which have no wings, mate with the females and die where they were born. The females then fly off to repeat the cycle. Interestingly, the female has a special organ that controls whether the egg she lays is fertilized or not fertilized. Unfertilized eggs result in sons, while fertilized eggs result in daughters. This control works to balance out the population.

Though the jewel wasp is beautiful and has amazing abilities, it has taken over the necessary job of corpse disposal. Thankfully, in Christ, we can eagerly await that day when Christ, Who has overcome death, will return to give to all who believe in Him the blessings of eternal life.

Prayer: Lord, I thank you for earning salvation for me. I long for your return. Amen.

Ref: Jack Werren, "Genetic Invasion of the Insect Body Snatchers," *Natural History*, 6/94, p.36

The Golden Minute

Daniel 3:1a
"Nebuchadnezzar the king made an image of gold, whose height was threescore cubits, and the breadth thereof six cubits. . . ."

Why do 60 seconds make a minute? Why does an hour have 60 minutes? After all, there are no natural reasons for this. A year is a natural time measurement, based on the sun's position in the sky. Well, believe it or not, there may be connection between the length of an hour and the golden statue built by King Nebuchadnezzar that Daniel and his friends refused to worship.

King Nebuchadnezzar's statue was 60 cubits high and 6 cubits wide. This is probably because the Babylonians used a counting system built on the number 60 and many of their buildings are measured in units or sub-units of 60. By 1300 B.C., the Egyptians had divided the day into 12 hours of daylight and 12 hours of darkness. Under this system, the 12 daylight hours were longer in the summer than winter daylight hours. But this system worked reasonably well, since Egyptians kept track of the time with sundials. It is believed that this system, along with the base-60 method of counting was borrowed from the Babylonians. The designation of 12 hours appears to have come from the Babylonians, since 12 is a factor of 60. The system passed on to the Greeks, who gave it to the Romans. It is thought that by the 13th century A.D., when accurate mechanical clocks were invented, the hour was finally divided into 60 minutes.

That our method of accounting time goes back over 3,000 years illustrates that ancient man was no less advanced than modern man. Over 3,000 years ago, man was smart enough to invent a time accounting system that still serves us well today.

Prayer: Dear Father, help me use the time you give me to your glory. Amen.

Ref: *Odyssey*, Fall 1998, "Inventing Time, How on earth did we get a 60-minute hour?" p.6

Of Course It's a Horse!

Proverbs 21:31
"The horse is prepared against the day of battle, but safety is of the LORD."

Just about every science textbook in print for the last century presents a diagram of the evolution of the horse. The first picture in the illustration shows a rodent-like, four-toed creature labeled *Eohippus* followed by *Mesohippus*, *Merychippus* and finally *Equus,* the modern horse. The number of toes decreases gradually from three to one. This diagram has convinced millions that scientists have proven horse evolution.

The diagram is, as some evolutionary scientists have admitted, a complete deception. One evolutionary biologist said, "The family tree of the horse is beautiful and continuous only in the textbooks." Why? First, *Eohippus* is almost identical to the modern day rock badger. It has nothing to do with horses. In addition, one-toed and three-toed horse fossils have been found in the same layer, indicating that three-toed horses didn't evolve into one-toed horses. Worse yet for evolution, three-toed horses still exist. The reason that there are so many varieties of horse in the fossil record and alive today is because God gave them a huge genetic range. Horses can be anywhere from 16 inches to over six-and-a-half feet high and can have anywhere from 17 to 19 pairs of ribs!

Rather than trusting in evolution's fraudulent "proof" for horse evolution, we should trust in the sure Word of our loving Creator.

Prayer: Dear Father, I thank you for the blessings the horse has brought mankind, but let my trust always remain in you. Amen.

Ref: Jonathan Sarfati, "The non-evolution of The Horse," *Creation* 21(3) June-August, 1999, p. 28

The Rocks Bear Witness

Luke 19:40
"And he answered and said to them, 'I tell you that if these should hold their peace, the stones would immediately cry out.'"

We have always been told that it takes millions of years for muddy deposits to turn into rock. Those who believe in creation have always disagreed. It seems pretty clear to us that much of the sedimentary rock we see today was deposited during the Great Flood at the time of Noah, which was not millions of years ago. Creationists generally place the Flood at about 4,600 years ago. Until recently, evolutionists dismissed this date as impossible: sedimentary rocks simply could not form that fast. Well, now it seems that they spoke too soon.

While digging some trenches in a salt marsh, a team of sedimentologists found stony nodules in the mud. Further research on how the nodules formed revealed that a mud deposit can be transformed into a layer of sedimentary rock in as little as six months! They found that two bacteria are responsible for this. One species gets its energy from the sulphates in seawater. In the process, it produces hydrogen sulphide. The second species of bacteria can do the same thing. But if too much hydrogen sulphide is present, it can also change iron compounds so that they react with hydrogen sulphide and other salts. The result is stony lumps of iron sulphide and iron carbonate—sedimentary rock that is hardened quickly enough to fossilize any animal before it decays.

Science has now confirmed what creationists who believe the Bible have always suspected. There are natural processes that can form sedimentary rock within the limited timeframe allowed by a literal reading of Scriptural history.

Prayer: Father, I thank you that your entire creation bears witness to you. Amen.

Ref: Andy York, "Set in Stone," *New Scientist*, 19 September 1998, p.25

God's Mercy Is Sweet

Psalm 119:103
"How sweet are thy words unto my taste! yea, sweeter than honey to my mouth!"

We hear many warnings these days that tell us not to eat foods loaded with cholesterol or fat. It's easy to come to the conclusion that the curse of sin has turned the whole creation against us. But God is a merciful God. He has given us the ability to learn about many of the beneficial foods and medicines that He has created.

One good example of this is Xylitol, also known as birch sugar. Earlier research showed that birch sugar, used more in Europe than America as a sweetener, inhibits growth of bacteria that cause tooth decay. Researchers in Finland wondered if birch sugar inhibited any other bacteria. They gave one group of pre-school children sucrose-based chewing gum and a similar group birch-sugar gum five times a day. The group that used the birch sugar gum had only about half as many ear infections. Scientists aren't sure how the birch sugar reduces ear infections. They suspect that it somehow prevents the bacteria that cause the infections from attaching to cells in the mouth and ear. If this is what happens, there is an added bonus. Since the sugar does not attack the bacteria directly, the bacteria would be unlikely to develop any defense against it.

In His mercy, God has given us more than just solutions to our problems in this life. Through His Son, Jesus Christ, He grants us the forgiveness of sins and eternal life. And through His Word, He tells us what salvation means.

Prayer: I rejoice in you, dear Father, because you are merciful in all things. Your saving Word is sweeter than any honey. Amen.

Ref: N.S., "A sugar averts some ear infections," *Science News*, v.154, p.287, October 31, 1998

Birds Egg Evolution

Job 39:14-15
"Which leaveth her eggs in the earth, and warms them in the dust, And forgetteth that that a foot may crush them, or that a wild beast may break them."

The creation is literally filled with millions of what those who believe in evolution call "happy coincidences." But when you encounter millions of instances of what appears to be thoughtful design, the obvious conclusion is that there is a Designer. For example, consider the design of bird eggs.

The shape of the egg makes it strong. This strength comes in handy in a busy nest. Mom and dad are coming and going, and they turn the eggs periodically during incubation. But all eggs are not equally egg-shaped, and there is a pattern to their shapes. Birds like robins that build a nice, dish-shaped nest tend to lay eggs that are more round in shape. Screech owls lay their eggs at the bottom of a hole in a tree and also have round shaped eggs. Birds like the killdeer barely build any kind of nest and lay eggs on the ground where round eggs could roll away. For this reason, birds such as a Killdeer lay much more sharply pointed eggs, which are designed to pivot on their small end. Likewise, eggs that are laid where predators are not likely to see them are usually pale or solid in color, but eggs laid out in the open are camouflaged. Moreover, baby birds that hatch in protected nests, like the bluebird, tend to be naked, blind and helpless. But the unprotected killdeer hatchlings are ready to leave the nest within minutes of hatching.

All coincidences? It seems more scientific to say that here we have a few of the many fingerprints of our wise Creator!

Prayer: I praise You, Father, for how your glory is reflected in the creation. Amen.

Ref: Jim Williams, "Bird basics: egg size, color and shape," *Star Tribune*, July 29, 1999, p.8

Spiritual Health Promotes Physical Health

Exodus 20:8
"Remember the Sabbath day, to keep it holy."

Most Christians realize that going to church and worshiping in earnest keeps them spiritually healthier. Now, a new study by Duke University researchers backs up an earlier study that people who attend church at least once a week do have better health and live longer lives.

Researchers interviewed almost 4,000 people between the ages of 64 and 101 each year for seven years. They asked about general health, their social activities and alcohol and tobacco use. During the seven-year period, weekly churchgoers were 28 percent less likely to die than those who were not regular worshipers or did not worship at all. The head of the study cautioned that the study does not show that those going to church for health reasons will get better. Rather, he said that it shows that those who attend worship for religious reasons have better health and live longer. An earlier study in California tracked over 5,000 people ages 21 through 65 over 28 years, producing similar results. Other studies have shown that regular churchgoers tend to have lower blood pressure, less depression and stronger immune systems than those who don't worship regularly.

We should stress that attending church simply to gain better health is not a conclusion of these studies. These studies were concerned with finding possible correlations between attitude and general health. The results indicate a relationship between spiritual health and physical health.

Prayer: Dear Father, I thank you that you have provided for my spiritual health through your Son, Jesus Christ. Amen.

Ref: Gary D. Robertson, N. C. "Study Shows Elderly Who Go to Church Live Longer, Healthier," *The Christian News*, August 1, 1999, p.24

The Amazing Mosquito Hawk

Job 39:26
"Does the hawk fly by thy wisdom, and stretch her wings toward the south?"

Flight is a big problem for those who believe that we owe our existence to evolution. Birds, mammals, reptiles, insects, and even some fish fly or at least glide through the air in controlled flight. So many different creatures fly that evolutionists must say that flight evolved several different times. The dragonfly is among the best fliers in the animal kingdom.

The dragonfly can beat its four wings in unison or separately depending on the maneuver it wants to make. Dragonflies can fly at speeds up to 25 miles an hour and even faster. They can hover, take off backward and even make an unbanked turn.

The dragonfly eats small insects, including mosquitoes, earning it the nick-name "mosquito hawk." A dragonfly can see a gnat from three feet away, fly to it, capture it and return to its original position in a just over one second! One-third to one-half of its body mass is made up of flight muscles. Its two eyes have a total of 60,000 lenses and are situated so that its range of vision is nearly 360°.

Dragonflies not only appear in the fossil record fully formed, but in much greater variety than today. One fossilized dragonfly was the size of a crow! Even the United States Air Force has studied the dragonfly to learn how it flies.

The dragonfly is no product of natural selection. It is clearly a specially-designed creature whose Designer understands flight better than we do. This Designer is our Creator God.

Prayer: I thank you, dear Father, for the beauty and wonder of the dragonfly. You are truly to be glorified! In Jesus' Name. Amen.

Ref: Richard Conniff, "The Lord of Time," *Reader's Digest*, June 1999, p.142

Unborn Babies Protect Themselves

Psalm 139:13
"For thou hast possessed my reins: thou hast covered me in my mother's womb."

When someone receives a donated organ they must take powerful drugs to prevent rejection. That's because the body's immune system identifies the implanted organ as "not me." It then assumes that the invader is dangerous and must be destroyed. That brings us to a puzzling mystery that scientists have been trying to solve: Why doesn't a mother's body recognize her unborn baby as being foreign to her body? After all, carrying genetic information from the mother and father, the unborn baby is genetically unique.

Some scientists theorize that the placenta is a physical barrier between the baby and the mother's immune system. Others believe that the unborn baby somehow hides from the mother's immune system. A third theory is that the mother's immune system is somehow forced to tolerate the unborn child.

New research from scientists at the Medical College of Georgia in Augusta, Georgia, supports this third alternative. Researchers have shown that the placenta produces an enzyme (IDO) that works to suppress immune cells. In effect, the unborn child puts together just the right enzyme to keep mom's immune system from attacking it. Scientists say that this discovery may lead to new drugs to treat autoimmune diseases and organ rejection.

This system or a similar system had to be fully operational in the first creature for successful pregnancies and births. It could not gradually evolve. This fact and the wisdom of this system provides more evidence of an all-wise, all-powerful Creator of every form of life and all that exists.

Prayer: I thank You, Father, that you protected me in my mother's womb. Amen.

Ref: J.T., "Don't reject me, fetus tells Mom," *Science News*, v.154, p.152, September 5, 1998

Ant Impostor Glorifies God's Creativity

Psalm 104:24
"O LORD, how manifold are thy works! in wisdom hast thou made them all: the earth is full of thy riches. . . . "

Deep inside an ant colony, perhaps in a decaying log in northern Idaho or even the north woods of Minnesota, there are ants who think they are taking care of their only pupae. In addition to attending to their young, however, they are providing a dangerous ant predator with food and protection as it matures. The predator, called Microdon, matures into a fly-like creature that lives only long enough to mate and lay eggs.

How does Microdon get welcomed into the ant's nest? It folds itself in half and ends up looking just like an ant larva. When researchers exposed some of the fly larvae in an ant nest, the ants quickly rescued the folded-over Microdon as if they were ant larvae. During its first of three larval stages Microdon enters an ant cocoon and eats the contents. During its two later stages Microdon moves unchallenged about the nest, eating more ant larvae. This fact at first puzzled researchers, since ants communicate and identify each other through specialized chemical signals. Further research revealed that Microdon are actually able to perfectly mimic this chemical communication!

But perhaps the most fascinating features of Microdon are the odd structures found on the outside of its third-stage. These vary from individual to individual and may look like toadstools or flowers. Scientists are unsure of their purpose. But they do marvel over the rich variety of the shapes, which seem unnecessary. While such variety may be unnecessary, they and the other features of Microdon testify to the rich creativity of our Creator God.

Prayer: Thank you, dear Father, for the rich variety found in your Creation. Amen.

Ref: Gregory Paulson and Roger D. Akre, "A Fly in Ant's Clothing," Natural *History*, 1/94, p.56

God's Loving Ownership

Psalm 50:11
"I know all the fowls of the mountains: and the wild beasts of the field are mine."

Have you ever tried to get a bottle in a baby's mouth in a dark room? It would be handy if the baby had a glowing target on its mouth. God has given the Gouldian finch, a native of Australia, a unique solution to this problem.

The Gouldian finch sports bright, parrot-like colors. A single bird may carry feathers in up to six brilliant colors, including sky blue, dark blue, yellow, green, purple and black. The finch's face may be red, orange or black. Found mostly in the grasslands of northern Australia, the finches usually build their nests of loosely-woven-dried grass. These nests are usually built in a tree hollow, a termite mound or in the grass itself. Consequently, finches often find themselves having to feed their young in low light conditions. Knowing the behavior He built into these birds, God has provided a solution for this problem. You might put a nightlight in the baby's room so you can find her mouth at night. God has done about the same for the Gouldian finch. Their young have two pairs of iridescent nodules marking the margin of their beaks. These nodules don't generate light. Instead, they are highly efficient reflectors that glow purplish-blue in all but total darkness. Lest anyone question whether these nodules were specially designed to help the parents feed their young, we note that the nodules disappear as the young birds mature into adulthood.

God can claim ownership over all living things because He made them. Because He made you and me He has not only claimed ownership over us, but even though we ruined His perfect creation with our sin, He sent His Son to save us.

Prayer: Father, I praise you for your love and salvation in your Son. Amen.

Ref: Carl Wieland, "Purple Pearls of Creative Wisdom," *Creation*, 21(3) June-August

Pollen Sheds New Light on the Shroud of Turin

Matthew 12:39
"But he answered and said unto them, 'An evil and adulterous generation seeketh after a sign, and there shall no sign be given to it, but the sign of the prophet Jonas.'"

New evidence not only suggests that the Shroud of Turin dates to the first century, but also places it in the Jerusalem area. The Shroud of Turin is a 13-foot long, three-foot wide piece of cloth that many people believe was the burial cloth of Christ. It bears a negative image of what appears to be a man who was crucified. A small bit of the cloth was carbon-dated in 1988 and the results led many to believe that the cloth came from the Middle Ages.

Now a botanist from Hebrew University has announced that he has identified pollen, taken from the shroud, of two important species that are found together only in the Jerusalem area. Pollen from the same two species was also identified on the Sudarium of Oviedo, a smaller cloth which has traditionally been considered the burial face cloth of Jesus. Both are stained with type AB blood. This cloth has been in Oviedo, Spain, since 760 A.D. and is documented back to the first century. Both cloths have pollen from a thistle which some have suggested was used to fashion Jesus' crown of thorns. The other pollen found on both cloths is from a common Jerusalem tumbleweed, an image of which also appears on the Shroud of Turin. Both plants bloom in Jerusalem around the time of the Passover.

While this intriguing new evidence suggests both cloths are authentic, we must be careful. The Christian faith does not rise and fall on such evidence. We stand on the Gospel of salvation through the forgiveness that Jesus won for us.

Prayer: Father, help me live a life of faith that bears witness to Jesus Christ. Amen.

Ref: (RSN) *REPORTER*, August 1999, p.9, *Minnesota Christian Chronicle*, August 12, 1000, p.9

Is Pizza Good for You?

1 Corinthians 3:16
"Know ye not that ye are the temple of God, and that the Spirit of God dwelleth in you?"

Could pizza be good for you? Research has shown that tomatoes, especially cooked, have powerful health benefits. That's because tomatoes have a powerful antioxidant called lycopene, in addition to beta-carotene, other B vitamins, vitamin C and other important minerals. A four-ounce tomato can provide you with one-third the recommended daily amount of vitamin C.

A European study of 1,300 men suggested that those who consumed the most lycopene in their food cut their heart attack rate in half, compared to those who had less lycopene in their diet. Another study that lasted five years followed the eating habits of 48,000 men. Those who ate 10 servings a week of cooked tomatoes, in whatever form, had one-third the rate of prostate cancer of those who ate less than two servings a week. Cooking tomatoes, as you would for pizza and pasta sauce, makes five times more lycopene available to you than eating the same amount of raw tomatoes. That's because cooking breaks down the cell walls of the tomato, making the lycopene more available for digestion. (Watermelons and pink grapefruit also have lycopene, but only as much or less as cooked tomatoes.)

Our bodies are called temples of the Holy Spirit by Scripture. God has richly provided us with healthy foods to keep our temples healthy. Even pasta sauce and pizza offer us health benefits in addition to tasty dining.

Prayer: I thank you, heavenly Father that you have given me the means to take care of my temple. Help me to use them. Amen.

Ref: Mayo Clinic Health Letter, September 1998

Those Gifted Red Knots

Psalm 145:15
"The eyes of all wait upon thee; and thou givest them their meat in due season."

Shore birds like the oystercatcher search for buried mollusks by touch. They poke around in the sand, hoping to find a hard-shelled mollusk. But if you have ever tried to find something that was hiding where you couldn't see it, you know that this method of looking for something is not very efficient. But another shore bird, the red knot, seems to know just where to find its hidden food.

Scientists observed that the red knot, a type of sandpiper, was seven to eight times more efficient at finding buried food than it would be if it were randomly searching. Red knots, who also search for food by pushing their bills into the sand, did better than those birds who search by touch.

The answer didn't come until scientists looked at the red knot's bill under the microscope. On the top of the bill they found tiny pits. Inside the pits they found cells called Herbst corpuscles. Scientists knew that other shorebirds have organs similar to these corpuscles that are used to feel vibrations from wriggling prey. They theorized that the red knot's Herbst corpuscles sense pressure changes in the displaced water under the sand when a mollusk obstructs the water's flow. Next they tested captive birds who were trained to find mollusks in pails of sand. Scientists found that if the sand was dry, the red knots did not do very well. But when the sand was wet, the red knots could indeed find the hidden mollusks.

God's creative nature and divine wisdom has provided His living creatures with a beautiful variety of ways of making their living.

Prayer: I praise you, dear Father, for the variety you have given us. Amen.

Ref: S.M., "New hunting trick explains bird luck," *Science News*, v.154, p.107

Airplane Fingers

Genesis 1:20
"And God said, 'Let the waters bring forth abundantly the moving creature that hath life, and fowl that may fly above the earth in the open firmament of heaven.'"

Aerodynamics, the science of flight, is a highly complex science. This is because many complex forces are acting on anything in flight. These forces include the power available for flight and drag produced by the flying object. Each of these categories includes, many additional forces that depend on the shape of the flying object, the shape and length of the wings, the speed and the altitude. This is why, for example, high altitude planes have very long wings.

One critical force that has been under recent study is the turbulence that forms at the tips of the wings. The shorter the wing, the more energy consuming turbulence forms at the tip of the wing. Different wing designs have been tried to decrease this turbulence. Engineers have had some success reducing this turbulence with winglets. You may have seen these small vertical wings on the wingtips of some airplanes.

Swiss researchers have been studying vultures with the hope of finding a better solution to this problem because vultures have a relatively short wingspan that has proven to be surprisingly efficient. They discovered that this is because the feathers at the vultures' wing tips spread out. They then tested a wing with a finger-like cascade of blades at the end. Their new wing was more than four times more efficient than the average wing design in use today! It takes a great deal of faith in evolution to think that natural selection possesses such knowledge of aerodynamics. Clearly the vulture was designed by an intelligent Creator who understands aerodynamics even better than we do!

Prayer: I thank you, dear Father, that we can see your wisdom in the creation. Amen.

Ref: "Wingfingers," *Flying*/January 1999, p.108

Learn of God's Loving Provision

Job 12:7-8

"But ask now the beasts, and they shall teach thee; and the fowls of the air, and they shall tell thee: Or speak to the earth, and it shall teach thee; and the fishes of the sea shall declare unto thee."

A number of species of fish live, apparently comfortably, in Antarctic waters as cold as the freezing temperature of seawater, which is a couple of degrees colder than fresh water. This presents us with several mysteries. First, cold slows down the chemistry necessary for life. At these temperatures, life's chemistry all but stops. Second, from a creation perspective, how could fish created to live in a perfect, warm, comfortable world be able to live in an environment that makes life seemingly impossible? A team of biologists from Stanford University's Hopkins Marine Station has found some remarkable answers to these questions.

The biologists studied an enzyme found in many creatures that changes a compound called pyruvate into lactate within the muscles. In the Antarctic fish this enzyme appears to help this conversion take place at speeds usually only found in warmer creatures. It seems that the cold-water fish have a slight modification of this common enzyme that allows the enzyme to work more quickly, despite the cold. And yes, researchers have also found the same modified version of the enzyme in a South American warm water species. It appears that God's foreknowledge of what would become of His once perfect world, and His desire to provide for His creatures, led Him to give this special version of this enzyme to creatures of His choosing. If God provides so wisely for fish, will He not provide for you and me, for whom His only Son died?

Prayer: I thank and praise you, dear Father, for your love in providing all I need, including a Savior. In Jesus' Name. Amen.

Ref: C.Wu., "Fish enzyme flexes to adapt to the cold," *Science News*, v.154, p.183

The Family that Prays Together...

Genesis 2:24
"Therefore shall a man leave his father and his mother and shall be cleave unto his wife: and they shall be one flesh."

"The family that prays together, stays together." A new survey by the Barna Research Group indicates that this old saying is still true. The phone survey interviewed 1,512 Christians about their marriage history and their church membership.

The respondents were divided into three groups. Couples who belonged to different denominations made up 17 percent of the sample. Couples from different denominations who decided to become active in the same denomination after marriage made up another 14 percent of the sample. The rest of the married couples, 65 percent, belonged to the same denomination when they were engaged. The lowest divorce rate was among those of different denominations before marriage and who decided to become active in the same denomination after marriage. This group's divorce rate was only 6 percent. The divorce rate for those who were raised in and remained active in the same denomination was 14 percent. The divorce rate for those who remained in different denominations was 20 percent, which is still less than half the divorce rate for the general population.

Researchers stressed that shared beliefs were not the primary reason for the success of these marriages. Rather, interviews showed that the most important factor in the successful marriages was shared religious activities. This is yet another study that reveals practical benefits for the lives of active church members compared with atheists or inactive church members.

Prayer: I thank you, Lord, for the gift of marriage. Help me always to honor it. Amen.

Ref: Study:"Divorce Rate Higher for Inter-Church Couples," *The Christian News*, July 26, 1999, p.17

Human Speech Itself Glorifies God

Proverbs 17:7
"Excellent speech becometh not a fool: much less do lying lips a prince."

Several different animals communicate on a limited basis with one another. But human speech is unique, leaving those who believe in evolution perplexed. The very oldest human fossils show the bony structures needed to support speech. Evolutionists will admit, in a candid moment, that they have no idea how speech could have evolved. One modern researcher said they have only "inferences based on hunches."

Some scientists have observed that human beings come with the built-in ability to learn and speak. While this idea is not popular among evolutionists, it is supported by the unique structure of the human vocal tract. No other creature has anything like it. The human larynx is placed low in the throat. That placement creates a sound chamber that allows us to make language expressive. Moreover, the placement prevents us from breathing and eating or drinking at the same time. But we are not born that way. A newborn's larynx is placed higher up in the throat, allowing a baby to breathe and suckle at the same time. By the time a child is six—and has no need to suckle and breathe at the same time, but is learning language—the larynx has moved to its adult position.

This obviously designed arrangement in support of human speech presents only more problems for the evolutionist. But for those who believe in our Creator God, it is one more testimony of His wise handiwork.

Prayer: Dear Father, I thank and praise you for the gift of speech. Amen.

Ref: Roger Lewin, "Spreading the word," *New Scientist*, 5 December 1998, p. 46

Don't Stress Out Your Unborn Baby

1 Peter 5:6-7
*"Humble yourselves therefore under the mighty hand of God,
that he may exalt you in due time; Casting all your care upon
him, for he careth for you."*

We all instinctively know that worry and anxiety is not good for
us. Previous studies have shown that stress affects our health and makes
our immune systems less efficient. But somehow that knowledge doesn't
seem to be enough to keep us from fretting about the challenges and
problems of life.

A recent study conducted by researchers from the State
University of New York at Stony Brook examined the effect of a
mother's stress on the birth weight of her baby. They studied 130 women
between the ages of 18 and 42 years of age. This exhaustive study
examined causes of stress including daily feelings of anxiety, long-term
anxiety and financial stress. They also factored in 66 medical risk factors
associated with the birth weight of their child.

The researchers found that premature births were most likely
among women who faced both medical risks and emotional stress. In
fact, they concluded that emotional stress was more likely to produce
underweight births than just medical risks alone. Researchers admitted
that every mother-to-be faces some stress, often without harm to the
baby. But the key is how one deals with negative events in one's life.

We can deal constructively with stress by giving all our cares to
God. Because God has loved us so much that He even allowed His Son
to be sacrificed for our salvation, we know He will carry all our cares.

*Prayer: I thank you, dear Father, that you have loved me so
much. I cast all my cares upon you, for I know you love me.
Amen.*

Ref: B. Bower, "Anxiety weighs down pregnancies and births," *Science News*, v. 138, p.102

Not Every Gift has Value

Romans 6:23
"For the wages of sin is death, but the gift of God is eternal life through Christ Jesus our Lord."

Human bridegrooms are not the only males who give gifts to their intended. The birds and the flies do it, too. The bowerbird woos his potential mate by building a hut for her and decorating it with bright objects. A relative of the dance fly presents his intended with a dead insect that is part of its diet. The bigger the insect, the more likely the female will accept the attentions of the male courting her.

The dance fly itself presents his intended with a gift, although the gift he offers is essentially worthless. The courtship takes place early in the morning as the males swarm. Each courting male carries what looks like a balloon in its back legs. These shiny balloons are only a few millimeters in diameter. As far as we know, they are of no value to the female, being made of nothing more than the male's saliva. That may be why the females do not necessarily select the males with the largest balloon. Why did God build this behavior into the dance fly? Perhaps it reflects nothing more than the incredible creativity and variety that God built into His creation.

Or perhaps it was to remind us that not every gift has value so that we would be doubly thankful for His good gifts to us. And His greatest gift to us was the life, death and Resurrection of His Son, Jesus Christ.

> ***Prayer: I thank you, dear Father, for the gift of your Son and the salvation He has made possible for me. In Jesus' Name. Amen.***

Ref: S.M.," Nuptial balloons, Size doesn't matter," Science News, v. 155, p.267

What Does the Milky Way Say About Evolution?

Job 22:12
"Is not God in the height of heaven? And behold the height of the stars, how high are they!"

When we look at the size of the universe, it sometimes becomes easier to see evolution as credible. After all, when we see galaxies that are said to be billions of light years away, it seems possible that there really have been millions of years. But, in truth, the immensity of space is really not so friendly to evolution's claim that the universe, and the earth, are billions of years old.

The Milky Way Galaxy, like many galaxies, is a spiral galaxy. You have probably seen the illustrations showing spiral arms of stars wrapped around the bright center of our galaxy. The stars closest to the center of our galaxy rotate around the center more rapidly than the stars further out on the spiral arms. When these differences in speed are worked out, we discover that our galaxy, no less our Earth, cannot be billions of years old. If the Milky Way was that old, the spiral arms would have long ago become nothing more than a disc of stars. Evolutionary scientists recognize this problem, calling it "the winding up dilemma." But they have been unable to come up with an explanation that is satisfactory.

As we look at the creation for evidence of its age, the clues we have so far found can only give us a upper limit. They cannot give us an exact age. But even the size of the universe fails to support the idea that the creation is billions of years old. If instead, we piece together the Biblical genealogies, which are really ancient calendars, we arrive at an age for the creation of a little more than 6,000 years.

Prayer: Father, I praise and thank you that though you are great, you have loved me through your Son, Jesus Christ. Amen.

Ref: D. Russell Humphrey's, Ph.D., "Evidence for a Young World," *Creation Matters*, July/August 1999

The Key to Real Leadership

Proverbs 12:24
"The hand of the diligent shall bear rule: but the slothful shall be under tribute."

Wasp colonies in temperate zones usually wait until the end of summer to raise male young. That's because male wasps are typically useless consumers of food. They don't do anything worthwhile in the nest, nor do they help gather resources. Female paper wasps often practice what expert's call "male stuffing." When food is brought into the nest, a job carried out by females, other females will stuff their brothers headfirst into unused nest cells. In the typical nest, males have short lives.

The first exception to this pattern was recently discovered in Costa Rica. Researchers from the University of Washington in Seattle have discovered a species of wasp in which the males are not only in charge, but also, help out in the nest. The female wasps still gather the food, but the males demand and get all they want, even from the queen. But the males were also observed fanning an overheated nest to cool it. They helped remove water during a flood, and they helped take care of the larvae. The males of this species remain with their parent colony for an unusually long time for wasps. Yet, at any time, they can father a new colony. Other tropical species are now being studied to see if this behavior is unique.

The Bible teaches that proper authority is respected when it comes from a spirit of servant hood. This pattern seems to extend even to wasps. Modern human culture needs to be reminded of the principle that when males pitch in and help, their leadership will be appreciated and followed.

Prayer: I submit myself to you, Lord, because you served me with your life, death and Resurrection. Thank you. Amen.

Ref: S. Milius, "Male Insects Rule in a Tropical Society," *Science News,* v. 155, p.116

The Amphipod's Unique Escape

James 4:7
"Submit yourselves therefore to God. Resist the devil and he will flee from you."

The hunted are smart to stay upwind of the hunter. Many predators hunt using scent, at least in part. The same principle works in water. Many predators smell chemicals given off by their prey and use that scent to help them locate their prey. If you were predator in a stream you would be smart to stay downstream from your prey. The prey won't smell you. In addition, you might be able to smell him if he gets too close.

Amazingly, the little shrimp-like amphipod seems to be able to smell predators even downstream. Researchers from Lund University in Sweden have discovered that amphipods can even smell a brown trout downstream and try to avoid it.

Scientists designed an artificial stream that divided into two branches. Down one branch was a brown trout, while no predator waited down the other branch. Then amphipods were placed upstream of the branch and allowed to go downstream. When the amphipods came to the branch, they avoided the branch with the trout. To check whether the amphipods actually smelled scents from downstream, researchers put trout scents in the downstream water. This was enough to cause the amphipods to avoid that branch. Yet when trout were placed in one of the downstream branches in glass tubes, keeping their scent from entering the water, the amphipods did not respond to the sight of the trout.

We, too, are being hunted as prey by the devil and his forces. We would do well to learn how to sense where the devil lies in wait for us, and like the amphipods, flee from him.

Prayer: Dear Father, I confess that I have not fled the devil when I should have. Forgive me through Jesus Christ. Amen.

Ref: S.M., "Downstream trout swim but can't hide," *Science News*, v.154, p.91

How Small Can Life Be?

Job 9:10
"Which doeth great things past finding out, yea, and wonders without number."

How small can a living thing be and still be considered alive. Viruses are not considered living things because they cannot reproduce on their own. But now scientists have confirmed the existence of bacteria that are smaller than the largest viruses and yet can reproduce on their own. They call this amazing class of bacteria nanobacteria. They are less than 20 percent the size of normal bacteria.

The unusual nanobacteria have some unusual abilities. Normally, cells in the immune system detect a bacterium in the bloodstream, engulf it, and destroy it. But nanobacteria are able to trick immune cells into committing suicide. Luckily, nanobacteria reproduce very slowly. While the average bacteria reproduce once an hour, nanobacteria take three days to double their numbers. But they still do a lot of damage. Researchers have found that nanobacteria are apparently able to build a calcium shell around themselves. It appears that when they enter the bloodstream, they quickly migrate to the kidneys, where they begin building their shells. Researchers investigating whether nanobacteria cause painful kidney stones have reported testing 30 stones so far. Each stone showed antibodies that bind to nanobacteria. The good news is that if nanobacteria are a cause of kidney stones, they can be eradicated with antibiotics.

While scientists are unsure of how something so small can be alive, our wonderful Creator is able to fashion even these tiny, amazing creatures.

Prayer: I rejoice and praise you, dear Father for the wonder of your creation. Amen.

Ref: John Travis, "The Bacteria in the Stone," *Science News*, v. 154, p.75

Saliva Is Nothing to Spit At

Malachi 4:2
"But unto you that fear my name shall the Sun of Righteousness arise with healing in his wings; and ye shall go forth, and grow up as calves of the stall."

Someday, instead of having to give a blood sample during your physical exam, the doctor might just ask you for a saliva sample. Saliva is an amazing fluid. Besides helping us moisten and digest food, saliva is able to speed healing and fight bacteria, fungi and viruses. Saliva is chemically almost identical to the clear part of your blood. It even has, in lesser concentrations, the infectious organisms found in your blood.

Let's say, for example that you suddenly find yourself in a stressful situation. The level of the hormone cortisol will increase in your blood in response to the stress. Within 20 minutes, that increase will be evident in your saliva.

Someday, saliva tests may replace blood samples. In the United States, saliva tests have been approved to diagnose AIDS, illegal drugs, periodontal disease, alcohol and premature labor. If the hormone estriol rises in a woman's blood before 36 weeks of gestation, doctors know that the woman may go into labor prematurely. Saliva testing is also used to check hormone levels in women who are having a difficult time conceiving a child. Other countries have approved a saliva test for hepatitis B.

Saliva's remarkable abilities are a witness to a loving Creator. He knew that we would sin and thus bring disease into the world. So He gave us saliva that, not only defends us against infection but also helps us heal. But the greatest love He has shown us is in sending His Son to give us spiritual healing.

Prayer: I thank you, Lord, for all your good gifts, especially salvation. Amen.

Ref: Judy Foreman, 'The spitting image gains credibility," *Star Tribune*, September 12, 1999, p. E3

Modern Medicine Is Catching Up with the Bible

Proverbs 3:7-8
"Be not wise in thine own eyes; fear the LORD and depart from evil. It shall be health to thy navel, and marrow to thy bones."

For thousands of years the Bible has taught that there is a link between spiritual health and physical health. When rationalism came along, many sought to deny the spiritual aspect of humans. For them, the spiritual and the material have nothing to do with each other. Even medicine was, and still is, taught and practiced by many without any consideration of humans spirituality. In one recent survey, only 20 percent of doctors reported that spirituality and healing ever came up during their medical education.

According to several recent surveys, that rationalistic view is now beginning to change. These surveys include a 1996 poll of family physicians, a 1997 poll of HMO professionals and a 1987 survey of Americans. Now, 87 percent of the American public believes that prayer and other religious practices help in the treatment of people who are ill. Most surprising, 99 percent of all physicians polled believe these things help. Forty-one percent of the American public polled said that their ill health had been improved or even cured because of personal prayer. It is not out of place, according to 74 percent of Americans, for doctors to begin a discussion of a patient's spiritual needs as part of the patient's treatment.

It's not surprising to learn that the Bible was right thousands of years before modern medicine learned that spiritual health and physical health are related. But it is nice to see this area of science finally catching up with the Bible.

Prayer: I praise you dear Father, because you have shared your wisdom with us in the Bible. In Jesus' name. Amen.

Ref: *Better Health*, Summer 99, v.15, n.2

The Deception of a Glow-in-the-Dark Shark

Jeremiah 17:9
"The heart is deceitful above all things, and desperately wicked; who can know it?"

The creation is filled with plants and animals that use deception to survive. The glow-in-the-dark cookie-cutter shark is a master at such deception.

The cookie-cutter shark is only a foot to a foot-and-a-half long. It doesn't have powerful muscles and can't swim very fast. But it doesn't need to. Lunch serves itself. The shark typically lives between 600 and 3,000 feet below the surface of the water. Looking down from above one of these sharks, you would probably not see it, because the top of its body is as dark as the deep waters where it lives. Looking up from below, you wouldn't see the dark outline of the shark against the light that filters down from the surface. This is because of the shark's first deception: its underside is completely covered with light-emitting cells that match the illumination from above.

The deception doesn't stop there, however. A small patch beneath the shark's jaw doesn't glow. From beneath, the cookie-cutter shark looks like a small fish, just the kind of dinner a tuna might like. As the tuna speeds toward what it thinks is a small fish, at the last minute the shark turns and takes a bite out of the tuna. Its open, round jaw takes a plug of flesh out of the tuna, hence the name cookie cutter shark. Interestingly, most cookie-cutter shark wounds are not fatal.

The fact that the cookie-cutter shark's form, deceptions and habits all match each other perfectly is not a result of chance, mindless evolution, but of God's careful design.

Prayer: Lord, enlighten me with your gifts that I may not be deceived. Amen.

Ref: S. Milius, "Glow-in-the-dark shark has killer smudge," *Science News*, v. 154, p.70

Rare and Hidden Flowers

Luke 11:33
"No man, when he hath lighted a candle, putteth it in a secret place neither under a bushel, but on a candlestick, that they which come in may see the light."

Two strange species of orchids are among the rarest flowers in the world. These unconventional plants live their entire lives underground, even flowering underground. Most often, their flowers never break the surface of the ground. One of these species, *Cryptanthemus*, has been found only three times in this century. Only 150 flowering *Rhizanthella* plants have ever been found.

Since they live in the dark underground, neither orchid has any need for chlorophyll or leaves. Each looks like nothing more than a white, waxy root. *Rhizanthella* produces a flower that is initially white, but which turns purplish as it ages. Its fragrance has been described as delicate and sweet. *Cryptanthemus* produces a candelabra flower stalk that has several white flowers with small red splotches. The flowers remain almost an inch under the soil. Both species begin life in the same way as most orchids. The orchid seeds germinate when threads of a fungus invade the embryo and form coils within the seed. At this point, other orchids grow leaves so they can make their own food. But these underground orchids continue to live in a symbiotic relationship throughout their entire lives, being fed by the fungus.

The wonders of God's creative handiwork never cease to amaze us. But God's greatest creation is a redeemed believer in Jesus Christ, a creation He does not want hidden. That's why He tells us in His Word to let the light of faith and new life He has given us in Christ shine onto those around us.

Prayer: Thank you, Father, for the new life you have given me through Christ and help me live so that others can see it. Amen.

Ref: Margaret Helder, "Underground Flowers in Australia," *Dialogue*, Winter, p.6

More About Amazing Aspirin

Psalm 147:3
"He healeth the broken in heart, and bindeth up their wounds."

A few years ago a "Creation Moments" program aired on how plants seem to have pain and alarm responses similar to humans and animals. Of course, we don't know if plants have feelings, as we know them. But when a plant is injured, it produces a chemical called jasmonic acid. This acid produces a vapor, similar to the jasmine in commercial perfumes, that is sensed by surrounding plants. In turn, the surrounding plants respond to the signal.

When humans feel pain, it is due to chemicals totally unrelated to jasmonic acids. As we all know, aspirin is important to pain management for many people. Aspirin works by disabling the chemicals that cause our pain. Now scientists have discovered that aspirin also works to shut down a plant's response to injury. Aspirin shuts off the plant's production of jasmonic acid, even though jasmonic acid is not at all similar to human pain-causing chemicals. More amazingly, aspirin's chemical reaction is of the same kind in plants and humans. So the next time you accidentally injure your favorite houseplant, it might appreciate a small dose of aspirin!

While we can be thankful that God placed substances in the creation which help us manage pain, we should never forget that nothing in this creation can deal with the underlying cause of pain. Only Jesus Christ can bring healing from our sinful condition that results in both spiritual and physical pain.

Prayer: Lord, I thank you because you have loved me enough to endure the pain of the cross to remove my spiritual pain. Amen.

Ref: S.M., "Aspirin works on plants, too," *Science News*, v. 154, p. 106

Is This Any Way to Do Business?

Mark 8:36
"For what shall it profit a man if he shall gain the whole world, and lose his own soul?"

Evolutionary scientists tell us that mutations were central to the production of the great variety of creatures on Earth today from the first living cell. At the same time, biologists admit that 99 percent of all mutations are harmful. But, argue evolutionists, if you add up that one percent of good mutations over millions of years, you get the beautiful variety of living things we have today. Let's take a closer look; even granting the unproven claim that one percent of mutations are helpful.

For augment's sake, let's compare evolution's business with human business. Let's say you start a business with $100. As is not unusual, you lose $10 the first year, the equivalent of a bad mutation. As you become more experienced in business, you lose less in the years that follow, but since only one out of 99 mutations is good, you continue to lose an average of $1 for the next 98 years. By the time your good mutation—your year of profit—comes along, you're $8 in debt. Even a big mutation - say a profit of $20 - will not be enough to make up for the next 99 years of even small losses. In short, this is no way to do business, either for you or for evolution. Neither you nor evolution can possibly make any gains, even with one percent of the mutations being good.

The evolutionary claim that all life today is a result of mutations runs counter to logic, common sense and simple arithmetic. The only way we can reasonably account for life's variety is to acknowledge the work of a wise and powerful Creator.

Prayer: Lord, help me to remember that all good things come from your hand. Amen.

The Nose Knows

Job 10:10-11
"Hast thou not poured me out as milk, and curdled me like cheese? Thou hast clothe me with skin and flesh, and hast fenced me with bones and sinews."

We all know that we each started out as a single cell in our mother's womb. Then those cells began to divide and within 21 days we had a working nervous system and a beating heart. How did all those new cells know how to hook up to each other to knit together our nerve connections? How do other cells know to hook up with each other to make our circulatory system, or form the heart? A fascinating new theory has now been proposed, based on the fact that every cell in your body has the same active genes that your nose uses to enable you to smell.

Known as olfactory genes, these genes have no known reason to be active in every one of your cells. Mammals and humans have more than a thousand different olfactory genes. One or another of these crisscrosses each of your cell seven times. Researchers already know that these genes were crucial to the development of your sense of smell as you formed in your mother's womb. As you formed, the nerve cells in your nose sent out growths, called axons, toward your brain. Each of those axons finally grew into the part of the brain called the olfactory bulb. Once there it hooked up with the cell designed to sense the specific scent detectable by the nerve cell in your nose that sent it out.

It has now been proposed that olfactory genes are active in every one of our cells because this same method of wiring our noses for smell is used to link all our cells as we form in the womb. If so, this system of knitting our unborn bodies together is an elegant and precise product of a loving Creator. You weren't put together by chance or by mindless forces.

Prayer: Father, I thank you that you hand made me, and still love me. Amen.

Ref: John Travis, "Dialing up an Embryo," *Science News*, v.154, p.106

Is the Mammoth Making a Comeback?

Psalm 50:11
"I know all the fowls of the mountains: and the wild beasts of the field are mine."

Will the mammoth walk the earth again? At one time, herds of these giant creatures, which stand 10 to 12 feet tall at the shoulder, covered North America, Europe, Asia and Africa. In late 1999, Russian and American scientists excavated a buried, frozen mammoth in Northern Siberia. Once excavated, the mammoth was airlifted, using Russia's largest helicopter, to Siberian ice caves that were fashioned into a laboratory. There the frozen mammoth was studied to see if either sperm, or DNA could be extracted from it.

This is not as fantastic as it sounds. A mammoth calf that was found in 1977 was found to have intact red blood cells. In 1978, DNA was first extracted from another mammoth. Researchers hope that they can extract intact sperm from the male they are now studying and use it to fertilize an elephant's egg. The result would be half mammoth and half elephant. If scientists can extract intact DNA another possibility opens up. The DNA can be inserted in an emptied elephant egg and the egg induced to divide. The result of this procedure would be pure mammoth of the *Mammuthus* species, one of six or seven known species. Scientists are able to use elephants in this work because there is only a five percent genetic difference between the two creatures.

If mammoths again walk among us, these huge creatures will glorify the God Who originally created them. They will add to the diversity of life we see around us that glorifies God's power and creativity.

Prayer: I thank You, Father, that you own me because of Christ's sacrifice. Amen.

Ref: Barb Williamson, "Agenbroad heads for Siberian woolly mammoth," (Online)September 20, 1999

93

Parrot Denies Evolution

Genesis 7:23a
"And every living substance was destroyed which was on the face of the ground: both man and cattle, and the creeping things and fowl of the heaven."

According to evolution theory, birds should not have been around at the time of the dinosaurs. This is especially true of the parrot, which is supposed, by those who believe in evolution, to be a more highly evolved bird.

When a fossilized parrot's beak was dug up 40 years ago, it was ignored. But was recently rediscovered at the University of California, Berkeley, by a graduate student. The problem, for evolutionists, is that they date the rock in which the beak was found as coming from the Cretaceous period when the dinosaurs lived and birds had not yet evolved. X-ray study of the fossilized beak shows that the beak has the same blood vessel and nerve channels as a modern parrot. But it is not just one parrot's beak that has been found in rocks from the age of dinosaurs. Loons, frigate-birds and other shore bird fossils have also been found in rock that was supposedly laid down during the time of the dinosaurs!

Fossils are remains of living things rapidly buried, we believe, at the time of the Genesis Flood. However, it is perhaps not too surprising that bird fossils and dinosaur fossils are generally not found together; indeed, bird fossils in any part of the earth's strata are extremely rare. This is because while dinosaur bones are very robust, bird bones and beaks are extremely fragile and would not have survived the turmoil of the Genesis Flood. The very few that have been fossilized tell us of very rapid and deep burial that would be expected during the Flood.

Prayer: I thank you, Lord, for making your Word trustworthy in everything. Amen.

Where Does True Knowledge Come From?

1 Corinthians 1:25
"Because the foolishness of God is wiser than men, and the weakness of God is stronger than men."

Americans lead the industrialized world in belief in creationism and a literal Bible. A recent study by a University of Cincinnati compares Americans' belief in a recent creation and the truth of the Bible with beliefs in other countries. The study found that 45 percent of Americans believe that God created the world less than 10,000 years ago. Another 40 percent try to mix God and evolution. Only ten percent believe in atheistic evolution.

In comparison, only seven percent of the citizens of Great Britain take the Genesis account of creation literally. Surveys of people in Norway, Russia and the Netherlands produced results similar to Great Britain. Interestingly five percent of the physical scientists polled in the United States believe the account of creation as taught in the Bible.

The study concluded that America is losing its scientific competitiveness because so many Americans believe in creation and the truth of the Bible. But scientific progress has come about because historically, scientists believed that we live in an ordered creation that follows the rules imposed on it by a Creator. They saw science as the discovery of God's order. Isaac Newton wrote more Bible commentaries than he did scientific books. Taking the Bible literally, he concluded that the earth was less than 7,000 years old. Other examples from history could be multiplied. But the lesson of human history is that true scientific progress is possible only when man recognizes that all knowledge and wisdom comes from our Creator God though the Scriptures He has given us.

Prayer: Dear Father, help me never to be deceived by man's false wisdom. Help me to rejoice in the knowledge you give. Amen.

Ref: "Americans for Creationism," *The Christian News*, December 21, 1998, p.18

There Could Be Magic Dragons

Psalm 148:7, 10
*"Praise the LORD from the earth, ye dragons and all the deep . .
. beasts and all cattle; creeping things and flying fowl. . . . "*

Numerous television programs have been made about the Komado dragon and the creatures are even going on display at some zoos. Reaching up to 400 pounds, they can even outrun a man. And they eat anything that they find dead or alive. Even a simple bite from a dragon will prove fatal within 72 hours unless you get immediate treatment. That's because its saliva contains 52 different deadly strains of bacteria that will produce blood poisoning. Its jagged, inch-long teeth make a tearing wound that ensures that plenty of poison will be delivered to your system. Even if you were bitten and escape, the Komodo dragon will be able to smell you from two miles away when its poison takes effect. These facts may lead some to ask not only why God made such a creature, but also why He has preserved them to this day?

Besides demonstrating variety in God's creation, the Komodo dragon may eventually save more lives than it has ever taken. When the dragon eats, its teeth tear its own gums, exposing its blood to those toxic bacteria. But the dragon never gets infected. Scientists have learned why. They have identified a protein molecule in its blood stream that kills all those toxic bacteria in its saliva. That protein is now being tested on mice. It could well be that the Komodo dragon has shown us the path to a super antibiotic that could treat the blood poisoning that kills hundreds of thousands of people every year. If so, this creature will again glorify God by giving us an important medical advance.

***Prayer: I thank you, Lord, that you have graciously placed
secrets in so many creatures that can help us in this life. Amen.***

Ref: Brian Eads, "Last Stand of the Komodo Dragon," *Reader's Digest*, August 1999, p.80

The Origin of Racism

Acts 17:26a
"And hath made of one blood all nations of men to dwell on all the face of the earth. . . ."

Does God approve of interracial marriage? To answer this question, we must look at the concept of race. The Bible teaches that we are all descended from the man Adam. Scripture also states we are all of one blood; Scripture never even uses the idea of race.

The descendants of Ham, who were cursed, were the Canaanites. Yet when Rahab, a Canaanite, came to faith in the true God, she not only was welcomed to marry a believer, but God included her in the line leading to Christ. The idea of different races, as distinct from different religions, was not much of an issue until 1859 when Charles Darwin published his famous book, *On the Origin of Species.* Darwin was a product of Victorian times and extremely racist in his views, always referring to colored peoples as "savages." Among the book's purported scientific claims for evolution was the claim that there are different races because some groups are more evolved than others. As this idea became accepted both in and out of the Church, racism became institutionalized. Today we know that typically the genetic differences between you and anyone else is only 0.2 percent.

Scientifically, there is only one human race, as Scripture clearly teaches. The Church can only combat racism by proclaiming the truth that all people on earth are one flesh, descended from one, real Adam, whose blood we share. It can also, proclaim the Gospel that all believers are spiritual descendants of the Second Adam, the Lord Jesus Christ, Who has redeemed us and made us new creatures.

Prayer: Help me, Father, love all people as your Son did when He died for them. Amen.

Ref: Ken Ham, "Inter-racial marriage: is it biblical?, *"Creation* 21(3) June-August, 1999

He Literally Walks on Water

Matthew 14:25
"Now in the fourth watch of the night Jesus went unto them, walking on the sea."

Would you like to be able to walk on water the way Jesus did? The closest human beings have come to walking on water is water skiing, which some even do without skis. But they still need a powerful and noisy powerboat to do it. But there is an animal that can walk on water without help of any kind.

The basilisk lizard is yet another tribute to God's unlimited creativity, knowledge and power. Basilisk lizards are found from Mexico to Ecuador and are equally at home in the trees, on the ground and in and on the water. The basilisk is an excellent climber and can outrun any enemy on the ground. Since its back legs are proportionately longer than its front legs, when it needs extra speed, it can run *just* on its back legs. It runs so fast that it can run across the surface of a body of water that's up to 100 feet wide. That's impressive for a creature that can reach two feet in length! Skin flaps on both sides of its back toes, as well as scale fringes on the toes, help the basilisk to stay on the surface. Should the basilisk run into danger while crossing the water, it sinks beneath the surface and either swims or runs along the bottom in a different direction. The lizard is able to stay underwater for several hours.

While the basilisk's method of walking on water is different from our Lord's, its abilities come from the same God. This lizard shows the wisdom and power of God to do things that seem impossible to us, including giving forgiveness of sins and eternal life to us through His Son Jesus Christ.

Prayer: I thank you, Father, because your wisdom is used in love for us. Amen.

Eight-Legged Thievery

John 10:10
"The thief cometh not but for to steal, and to kill, and to destroy. I am come that they might have life, and that they might have it more abundantly."

Out of more than 30,000 species of spiders, there are three species that make their living through thievery. That number was recently raised from two with the discovery of a tiny, thieving species on the island of Taiwan. The other two species live in South America and are known to steal only silk from the webs of other spiders.

This strange little spider from Taiwan, which steals much more than just silk, is described as looking like a tiny drop of mercury with legs attached. It preys on the webs of giant wood spiders, whose webs can be more than three feet across. Many spiders eat their own webs for the purpose of recycling the silk to repair a tattered web. But this newly discovered little thief eats the webs of the giant wood spider. Researchers say that these little spiders can, on the average, make a wood spider's web 21 percent smaller. They have also been seen stealing small insects from the wood spiders' web as well as wood spider eggs. They are able to do this because they are small and rather sneaky. They do spin their own webs when they are ready to lay eggs, and they also spin their own silk to cover their eggs.

While it is wrong for us to steal, theft is apparently the way God has chosen to feed these spiders in a fallen world. That our world is corrupted by our own sin should remind us that God is our ultimate provider. He has even prepared forgiveness and eternal life for us through the holy life and innocent suffering and death of His own Son, Jesus Christ.

Prayer: Dear Father, help me to keep your commandments because you have given me salvation in Jesus Christ. Amen.

Ref: S. Milius, "New spider: Unusual suspect steals web," *Science News*, v. 154, p. 53, "More than 30,00 species, "Spiders" *Grolier*, 1995, Multimedia Encyclopedia

B. C. - Before Columbus

1 Kings 9:26
"And king Solomon made a navy of ships at Eziongeber, which is near Eloth on the shore of the Red Sea, in the land of Edom."

Normally, it would not be unusual to find a stone with engraving on it in South America. But a stone found in 1872 in Brazil is unusual since Aztec, Mayan and Incan civilizations never lived in that region. Besides, the engraving on the stone that was found in 1872 was in Phoenician. Its discoverer, who did not know Phoenician, sent it to authorities who gave it to the head of the national museum. He recognized the ancient language and translated it. It reads, "We are the Sons of Canaan from Sidon, the city of the King. Commerce has cast us on this distant shore, a land of mountains."

Could 7th century B.C. Phoenicians have sailed as far as South America in search of commerce? In the ancient world, Phoenicians were known as the greatest commercial sailors of their day. But there is a modern bias against seeing ancient peoples as being as smart and resourceful as we are today, so the inscription has generally been considered a hoax. Then, in the 1960s, the original rubbing made of the inscription turned up at a scholarly auction in the United States. A modern expert on ancient languages studied the rubbing and concluded that the inscription must be authentic. It contains quirks of the Phoenician language that were unknown even to scholars in the 19th century.

While we today may know more facts, the ancients were no less intelligent and resourceful than we are today. That's consistent with the Bible's picture of man as God's highest visible creation, rather than evolution's gradually improving creature.

Prayer: I glorify your Name, O Lord, for making us the way you have. Amen.

Ref: Archeology, "Before Columbus or the Vikings," *Science*, May 1968

Are You Growing New Brain Cells?

Psalm 107:8
"Oh, that men would praise to the LORD for his goodness, and for his wonderful works to the children of men!"

It appears that we are made even better than we thought. The human body and its systems are engineering wonders. The human brain is the most complex arrangement of matter in the universe. Now, new research shows that your brain is even more wonderfully made than anyone thought.

It used to be believed that the brain cells you were born with were all the brain cells you would ever have. Then, research showed that canaries grow new brain cells to learn new songs. Later research showed that the human hippocampus grows new cells. That's the part of your brain where you form memories and remember new faces. The latest research now shows that your cerebral cortex also gets new cells every day. That's the part of your brain where you do higher intellectual functions and where your personality lies. The study, which used macaque monkeys, concluded that literally thousands of new brain cells are added to your cerebral cortex every day. Scientists concluded that this also happens in humans because monkeys' bodies usually work pretty much the same as human bodies. Scientists hope that this finding may help develop new treatments for degenerative brain diseases and to replace cells that are lost with age.

God has given us wonderful bodies that have amazing abilities, many of which we are still learning about. There is not enough good luck in the universe to enable blind evolutionary chance to create our bodies, much less our brains. Our very existence bears witness to God's wisdom and power.

Prayer: I give thanks to you, Lord, for all your gifts that make me possible. Amen.

Ref: "Study finds that new brain cells arrive daily," *Star Tribune*, October 15, 1999, page A20

The Armored Virus

Ephesians 6:11
"Put on the whole armor of God, that ye may be able to stand against the wiles of the devil."

In Darwin's day, everyone thought that single-celled microbes were simply bags filled with some kind of jelly. Today, of course, we know that even a single-celled creature is a complex living system that contains miniaturized organs to carry on all the business of life. A new discovery now shows us that some viruses are even more elegantly complex than we thought.

Viruses that infect bacteria, called bacteriophages, inject their DNA into bacterial cells with something that looks like a syringe. Some bacteriophages have been found to have a unique outer shell. Bacteriophages form their outer shells, called capsids, by making hundreds of copies of the same protein. This is not a simple process. A bacteriophage known as HK97 goes through a multi-step process in which it assembles five or six proteins into a ring. These are then interlocked, like chain mail, resulting in an outer shell that, like a gemstone, has 72 faces. This structure results in a strong outer armor. But it requires that the protein be folded and assembled and then changed into this interlocking pattern. As one scientist put it, "People don't know how to make proteins, let alone fold, assemble, and then transform them into something like this."

Contrary to what Darwin's followers still believe, there is no simple form of life. That's because there is a Creator, and He will provide all who follow Him with salvation and then He will give us spiritual armor to protect us from our spiritual enemies.

Prayer: Father, I thank you for the beauty and love that you give to us. Amen.

Ref: "Protein Chain mail offers armor for viruses," *Science News*, Vol. 154, July 18, 1998, p.38

Church Attendance Can Result in a Long, Happy Life

Hebrews 10:25
"Not forsaking the assembling of ourselves together, as the manner of some is, but exhorting one another, and so much the more as ye see the day approaching."

Studies show that people who attend church regularly tend to be healthier than those who don't go to church. Now, a new study shows that those who attend church regularly also live longer than who don't. Regular church attendance, as in the other studies, was defined as going to church at least once a week. This new study was done by the Population Research Center at the University of Texas at Austin.

The data for the study began to be collected in 1987. Researchers from the Center for Disease Control interviewed 22,000 people in their homes about cancer risk factors. Then researchers from the Population Research Center studied 2,000 of those same people who had died between the original interviews in 1987 and 1995. They concluded that, on the average, those who attended church at least once a week lived seven years longer than those who didn't. Those who never attended church lived, on the average, to be 75 years old. But those who were regular in church attendance lived to an average age of 82. Researchers say that those who were regular in their church attendance may benefit from following their church's advice to avoid unhealthy behaviors. Further, they say, the social ties developed by regular church attenders result in close relationships with others who help monitor their health. Of course, the study could not deal with the spiritual benefits of going to church. We learn those from reading about all the gifts God promises us in His Word.

Prayer: I thank you, Lord, for all the gifts I receive from you through your Word. Please bless and help my pastor. Amen.

Ref: David Briggs, "Study Reveals Churchgoers Live Longer," *Christian News*, December 21, 1998

The Molecular Computer

Scientists are recognizing the similarities between the workings of the DNA code and computer science. DNA is, after all, a complex information storage system. But comparing DNA as an information storage system to our largest and fastest computers is like comparing the Space Shuttle with a bow and arrow.

Harvard biologists have been amazed at what they have learned about the DNA workings of a single-celled protozoan. Its DNA routinely solves mathematical problems that only a very advanced modern computer can handle.

All of us have seen the complex problem of the traveling salesman. He has seven points of departure, and seven destinations. The problem is to work out the most efficient route for him to follow. By comparison, the protozoan under study solves the equivalent problem with up to 50 points of departure and 50 destinations.

This little protozoan has two nuclei, with nucleus arranging the same DNA differently. The smaller nucleus stores only the DNA needed to live in small units and must be reassembled before they do anything. This means it may discard up to 95 percent of the DNA, similar to compressed computer information. When the protozoan needs to reassemble its DNA, each end of each unit needs to find the proper end of the only other unit it attaches to.

This is not "simple" life. Even scientists now recognize that this one-celled creature is a complex computer. And this bears witness to God, Who is the author of life, mathematics and all knowledge.

Prayer: I thank you, Father, that you have given knowledge to us. Amen.

Ref: Kathy Sawyer, "Biological Software," *Princeton Alumni Weekly*, June 10, 1998, p.7

Chaco Canyon Mysteries and Biblical History

2 Timothy 3:16
"All Scripture is given by inspiration of God, and is profitable for doctrine, for reproof, for correction, for instruction in righteousness..."

What do we really know about human history? Much of ancient history was recorded by scribes who worked for the king. Their job was not necessarily to record history as it took place, but to make the king look good. Much of the West's history has, for over a century, viewed man as traveling from a primitive state to a much more sophisticated and technological state. But actual history challenges this view.

Take, for example, the Anasazi Indians, whose center of civilization seems to have been in the Chaco Canyon area of New Mexico. They are thought to have flourished about 1000 years ago and they built hundreds of miles of roads leading from all directions to the canyon. The roads are straight and always about 27 feet wide. The so-called North Road runs straight north for over 30 miles.

But even more impressive are the nine Great Houses they built. One was at least five stories tall, covered three acres, and had 650 rooms! These houses were built of millions of cut sandstone blocks. The floors of the Great Houses required the Anasazis to haul 215,000 trees—averaging almost 9 inches in diameter and weighing 600 pounds—50 miles from the forested mountains! Yet this sophisticated, hard-working civilization has disappeared without a trace.

Our knowledge of man's early history is not as complete as we are often led to believe. But we can trust the history that is presented in the Bible.

Prayer: Dear Father, I thank you for your trustworthy Word. Amen.

Ref: "The Chaco Canyon Road System," *Archaeoastronomy*, 4:50, October/December. 1981

Is the Earth Praising the Lord?

Psalm 96:11
"Let the heavens rejoice, and let the earth be glad; let the sea roar, and the fulness thereof."

Every bell made has its own tone, depending on its size, shape and composition. It turns out that the Earth, too, rings like a bell. This ringing, called free oscillation, is caused by earthquakes. A powerful earthquake in Chile in 1960 with a magnitude of 9.5 left the entire Earth ringing for days afterward. As the Earth rang, its entire surface rose and fell by a centimeter every 54 minutes. But it happened so slowly it could be detected only by sensitive instruments.

It was while studying these earthquake-caused oscillations that Japanese researchers discovered that the Earth also constantly vibrates. These vibrations appear to have nothing to do with earthquakes. The vibrations were discovered using an extremely sensitive instrument that measures gravity. These constant vibrations, with periods ranging from two to eight minutes raise and lower the surface of the Earth the diameter of a single molecule. Scientists at the University of California, Berkeley, are now trying to confirm the vibrations.

What's causing this constant vibration? There is so much energy in these vibrations that they cannot be accounted for by small earthquakes. Scientists are theorizing that it could be caused by the earth's winds. They say that there is enough energy in the Earth's winds to cause the vibrations. Others speculate that there could be some unknown processes going on within the Earth that may be causing the vibration.

Or, we might suggest, scientists have discovered that the Earth does literally praise its Maker through these low-frequency vibrations.

Prayer: Dear Lord, I praise and honor you with the rest of your handiwork. Amen.

Ref: Richard Monasterksy, "Ringing Earth's Bell," *Science News*, v.154, July 4, 1998, p.12

Evolving View of Neanderthal Doesn't Help Evolution

Genesis 4:20
"And Adah bore Jabal: he was the father of such that dwell in tents and of such as have cattle."

The first Neanderthal was discovered in 1856. Evolutionists quickly embraced it as evidence of human evolution. To promote Neanderthal as proof of human evolution they, reconstructed him to look like a human-ape cross. As more evidence accumulated over the years, Neanderthal began looking more human. His posture began to straighten, and his hair started disappearing. The look on his face even became more intelligent and aware.

Today there are several schools of thought about Neanderthal man. While some hate to lose what they consider to be evidence of human evolution, others are saying that new evidence is making Neanderthal more human than ever.

In 1997 German researchers found three aerodynamic wooden spears that are credited to the Neanderthals. Other researchers in France and Portugal have discovered what they call rather sophisticated ornaments and tools in a cave where Neanderthals lived. Those who want to hold to the older view about Neanderthals say that Neanderthals simply imitated humans who made jewelry. But the French and Portuguese ornaments are unique in style. Pendants have also been found at Neanderthal sites, which, researchers say, indicate that Neanderthals understood symbolism as well as we do. Older views of Neanderthals as scavengers are now challenged by evidence that they hunted and butchered their own food.

Sure, Neanderthals lived in caves, but so have other humans. Researchers are concluding that Neanderthals were just as human as we are, as we would expect from the biblical account of creation.

Prayer: I thank you, Lord, for a place to live. Protect the homes of your people. Amen.

Ref: Jeffrey Brainard, "Giving Neanderthals Their Due," *Science News*, v.154, August 1, 1998, p.72

Loving Plant Mothers

Genesis 1:11
"And God said, `Let the earth bring forth grass, the herb yielding seed, and the fruit tree yielding fruit after his kind, whose seed is in itself up on the earth'; and it was so."

Different kinds of creatures have different ways of taking care of their children. It is the father sea horse who carries the eggs until they hatch. Mother ostriches totally ignore their eggs and hatchlings, leaving all the care to father ostrich. While these examples may be odd, we would normally think it even stranger if some mother plants cared for their young. Yet two different species of plants have been found where the mother plant does indeed help her seedlings survive.

Both plants live in the very harsh setting of cold, dry mountain slopes of western North America. One of the plants is a thistle; the other is the monument plant. Both plants live for decades, growing only a little every year. But the mother plant is saving energy and storing water. Then, at flowering time, both plants send up a tall flower stalk. Once the seeds are ripe, the plant falls over dead. The seeds are not dispersed, but hidden in mom's vegetation. As mom decays, the seeds begin to germinate, using all the water that the mother plant saved up.

In addition, as they grow under the protection of mom's remains, the seedlings will not have to compete with mom. Scientists who studied seed germination rates of both plants concluded that the mother plant's saved water provides a big advantage to the seedlings. Protected monument seedlings survived twice as well as those without protection. Four times as many thistle seedlings survived when protected. God cares for all His creatures, and mothers are one means He uses to take care of our needs when we're young. That's even true for some plants!

Prayer: Father, I thank you for my mother. In Jesus' Name. Amen.

Ref: S. Milius, "Parental Care Seen in Mountain Plants," *Science News*, v. 154, July 11, 1998, p.20

The Miatsos' Good Memory Supports the Bible

Genesis 10:1
"Now these are the generations of the sons of Noah: Shem, Ham, and Japheth: and unto them were sons born after the flood."

Some people think that Noah's flood was a local event in the ancient Near East, rather than the world-wide catastrophe described in the Bible. Besides the fact that the local flood view rejects the Bible as the Word of God, there are many other problems with it as well. At least 138 cultures around the world have legends of a giant flood in which only a few people were saved in a large boat. Those who believe that the biblical flood was only local respond by suggesting that there have been many large, local floods all over the world.

Even that response fails to explain many things. Take, for example, the Miatso people of China. These ancient people are not Chinese or Semitic. When the first Christian missionaries visited the Miatso people, they were surprised to learn that the Miatso remembered not only the creation, but also Noah's flood quite accurately. According to their legends, the name of the first man created translates as "dirt." In Hebrew, Adam's name translates as "earth" or "clay." The important son of Adam is Seth in the Bible and Se-teh to the Miatso. The name of the man who built the big boat that saved all mankind is Nuah to the Miatso. According to their legends, he had three sons, Lo-Han, Lo-Shen and Jah-phu. Noah's three sons, of course, were Ham, Shem and Japheth. The Miatso say they are descended from Jah-phu.

If the great flood of Noah was merely a local, near Eastern event, how come the Miatso people remember the biblical Flood, the creation, and the names of the people involved?

Prayer: I thank you, dear Father, that your Word is trustworthy in everything. Amen.

Magnetic Turtles

Job 38:4
"'Where wast thou when I laid the foundations of the earth?
Declare, if thou hast understanding.'"

When it comes to sailing the seas, the compass is one of man's greatest inventions. That's why researchers were surprised to discover that some bacteria have tiny, built-in compasses. But these compasses are simple affairs based on tiny bits of iron within the bacteria. Now it has been discovered that loggerhead turtles not only have a built-in compass, but also use it to pinpoint their location.

Depending on position, the Earth's magnetic field encounters the earth's surface at differing angles. Using very sophisticated equipment, we can measure these angles and determine longitude and latitude. Young loggerhead turtles must stay within the ocean current system that surrounds the Sargasso Sea in the center of the Atlantic. When they get so far north in this current system that they might leave it, they turn south and return to a more central location within the system. When they get too far south, they turn north. To see if the turtles could actually sense their location based on magnetic fields, researchers tested loggerheads in tanks surrounded by electric coils. When they simulated magnetic conditions on the northern end of the turtles' boundary, the turtles swam south. When they simulated magnetic conditions at the southern boundary, the turtles swam north.

If the loggerheads' navigation system evolved slowly over millions of years, it would have been of no survival value to the turtles until it was complete. Since the turtles would have a very low survival rate without it, their navigation system bears witness to the all-wise, powerful Creator Who made all things.

Prayer: Dear Father, I praise you for all the wonders of your creation. Amen.

Ref: Jonathan Sarfati, "Turtles—reading magnetic maps," *Creation 21*, March-May 1999, p.30

The Probability of Evolution

Job 10:12
"Thou hast granted me life and favour, and thy visitation hath preserved my spirit."

Evolutionists explain that life started from nonliving matter in some sort of "primeval soup." Over the years they have presented several scenarios of how this might have happened. They have suggested isolated ponds, pools of water on the sides of volcanos, and in the oceans. Let's consider the general mathematical principles involved in any of these scenarios.

In science, the person who proposes a theory is supposed to present the evidence for that theory. Yet, for the incredible claim that life sprang into being out of non-living materials, no evidence is offered. Just how incredible are evolution's claims about the origin of life? Given the conditions evolutionists claim existed at the origin of life, the chance of evolving the simple, common, iso-1-cytochrome "c" protein is one chance out of one followed by 75 zeros. But that's not a living, reproducing thing. Given the same conditions, the chance of getting a DNA molecule with the ability to reproduce is 100 billion, billion to one. The chance of getting a minimal cell works out to one chance out of a 1 followed by 4,478,296 zeros!

Rigorous examination of evolution's claims about the origin of life shows that every evolutionary claim about how life started is just as fanciful. As every believer has testified, God is the source of all life, including yours and mine. It has always been so and will always be so.

Prayer: Dear Father in heaven, I thank you for the gift of life and eternal life. Amen.

Ref: Joseph Mastropaolo, "Evolution Is Biologically Impossible," *Impact* #137, November 1999

The Role of Fungus in the Life of Orchids

Psalm 96:3
"Declare his glory among the heathen, his wonders among all people."

Orchid seeds typically cannot begin life without a good fungal infection. The tiny, hard seeds cannot begin to sprout until fungal threads grow into the tiny embryo inside. This prompts sprouting because the fungus converts the nutrients stored in the seed to a form the embryo needs. Without that conversion, the seed will never sprout. The embryo cannot even absorb the water it needs without the help of the fungus.

By the time they have green leaves, most young orchids are able to make their own food and absorb their own water. When the orchid enters a dormant stage at the end of the growing season, its relationship with the fungus of its birth is forever over. But when the new growing season begins, the dormant orchid must again be infected by a fungus until new roots and stems form and grow new green leaves. Some orchids remain completely dependent on the fungus for nutrients, minerals and water.

The orchid's complete dependency on fungus to sprout or, in some cases, to provide all its nutrients and water, makes it a wonder that there are any orchids at all. If evolution were true, it is highly unlikely that the first orchid would have evolved where exactly the right fungus was present. A much simpler explanation is that the orchid and the fungus that infects it were intelligently designed by God for this special relationship.

Prayer: Dear Father, I cannot live without you and the salvation you have given me through your Son Jesus Christ. Amen.

Ref: Peter Bernhardt, *Wily violets & underground orchids*, p. 219

Did Ancient Man Make Artificial Basalt?

Genesis 4:22a
"And Zillah, she also bore Tubal-cain, an instructor of every artificer in brass and iron."

The fertile land around the second-millennium-B.C. city of Mashkanshapir provided little stone for building. Excavators of the city, which is in modern Iraq, believe they have discovered the ancient Mesopotamians' ingenious solution to this problem. They *made* rock. Their conclusion is based on the nature of what researchers first thought was natural basalt, which was evidently used to build many structures.

Basalt occurs naturally nowhere in the area. Furthermore, all the hundreds of basalt rocks found in the area are flat on one side and bumpy on the other. The chemical makeup of the rock is similar to the silt found in area rivers. The basalt was formed, say researchers, by forming the river sediment into slabs, melting the sediment at almost 2,200° F, and then allowing it to cool slowly. Researchers think that the development of such an unexpected and sophisticated process was a result of potters and metal smiths pooling their knowledge.

Traditionally, researchers have held to an evolutionary view of man that views ancient man as technologically unsophisticated. This view rejects the Bible's claim that we are intelligent beings, created by an intelligent God. But the man-made basalt of ancient Mesopotamia supports the Bible's claim that as early as the seventh generation after Adam, metalworking had developed into a sophisticated science.

Prayer: I thank you, Lord, for the intelligence you have given me. Help me to us it to glorify you. Amen.

Ref: B. Bower, "Ancient Mesopotamians made rock from silt," *Science News*, V. 153, June 27, 1998, p. 407

The Red-Eyed Sniper Fish

Genesis 1:20a
"And God said, `Let the waters bring forth abundantly the moving creature that hath life..."'

The fish known as the loose jaw or dragon fish has been found to have some remarkable abilities. At most 10 inches long, this deep-sea fish lives at depths between 1,500 and 4,500 feet. At these depths the only light that gets through the water is a dim blue light. As a result, sighted creatures that live at those depths can see well in dim blue light, but virtually none of them can see light in the red end of the spectrum. None, that is, but the dragon fish, and that's its secret weapon.

The dragon fish is the only known creature to have chlorophyll in its eyes. This compound gives the dragon fish the ability to see red light in the depths. But why would the dragon fish have these abilities if there is no red light to see? He makes his own red light. This bioluminescent red light is invisible to the dragon fish's prey. But the dragon fish can see its red light reflecting off nearby prey, as a sniper who uses an infrared scope to sight his prey in the dark.

The dragon fish offers a powerful set of arguments in favor of a Creator. Its unique ability to see red where there is no red light cannot be explained by natural selection. Before it supposedly evolved the ability to see red light, how would it know about red light? And why would it have evolved the ability to create a color of light it could not have known about? The dragon fish's sniper-like ability to find its prey with light invisible to its prey bears powerful witness to a wise and powerful Designer.

Prayer: Dear Father, I praise you for the wonders you have made. Amen.

Ref: S. Milus, "Red-flashing fish have chlorophyll eyes," *Science News*, V. 153, June 6, 1998, p. 359

The Largest Plant Cell Ever?

Genesis 1:28a
"And God blessed them, and God said unto them, "Be fruitful
and multiply, and replenish the earth and subdue it..."

One wouldn't think that algae would be very interesting. We've all seen the stuff wherever there is water. But researchers have recently discovered an algae that would make a good star in a horror movie. The algae known as Caulerpa taxifolia was first seen in 1984 off the coast of Monaco. It covered only about a square yard of sea bottom, but it covered it completely, smothering all other plants. Only six years later it had reached French waters. Four years after that, it had reached Spanish waters. It can grow on mud, rock or sand, and wherever it invades, it chokes out all other plants. A square yard of the algae can contain 700 feet of "stems" and 500 fronds!

Ordinary members of this algae species grow only about ten inches high. But this algae is huge and can have a "stem" that is nine feet long, with 200 fronds dangling from it. Moreover, each complete plant is a single cell. This could be the largest plant cell ever! Cutting it won't control it, because even a tiny bit of it will grow into a new plant. Researchers think that the original parent developed in an aquarium because all of these algae found so far turn out to be male. If true, this alga could become a classic example of how man has ignorantly modified the balance of God's well-designed creation to his own harm as well as to the harm of other creatures. That balanced inter-relationship between creatures is a witness to God's wise design.

Prayer: I thank you, Dear Father. You have given us a
well-designed world. Amen.

Ref: Janet Raloff, "Rogue Algae," *Science News*, v.154, July 4, 1998, p.8

The Secular Benefits of Christianity

Proverbs 11:11
"By the blessing of the upright the city is exalted, but it is overthrown by the mouth of the wicked."

The Bible teaches that believers bring God's blessings to even the unbelievers who live around them. For example, the Lord blessed Potiphar's entire household because Joseph was placed in charge of it. Yet enemies of Christianity, especially secular humanists, almost never grant that Christians bring blessings to those among whom they live.

Recently, Guenter Lewy of the University of Massachusetts set out to write a book entitled *Why America Doesn't Need Religion.* He wanted the book to be "a defense of secular humanism and ethical relativism." Lewy is not a Christian, and does not believe in God. But he was determined to offer his results as objectively as possible.

As Lewy assembled his extensive research he received a surprise. He found himself forced to conclude that Christianity has a record of strong support for social justice and human dignity. Other research forced him to conclude that Christians constantly show a lower rate than non-Christians of the behaviors associated with social ills and moral failure. These include divorce, domestic violence, out-of-wedlock births, adult crime and juvenile delinquency. He finally concluded, from other studies, that people who actually live the Christian life have higher rates of happiness and are healthier. The final title of his book is *Why America Needs Religion.*

Christians should not be fearful about living out their faith, even among unbelievers. As they live out their faith, God is not only blessing them, but also the unbelievers who surround them.

Prayer: I thank you, Dear Father, for all your goodness to me. Help me live out the faith you have given me. Amen.

Ref: Charles Colson, "The Gospel according to Jesse: Is religion a crutch?" *Minnesota Christian Chronicle*, December 2, 1999, p. 16

The Amazing Leaf-Miner

Mark 13:28
"Now learn a parable of the fig tree; When her branch is yet tender, and putteth forth leaves, ye know that summer is near."

If you live in a climate where fall brings colors to the leaves before they fall, you've probably noticed something strange. Once the leaves start to fall, the ground is littered with their color. But among all those fading leaves, you might find a perfectly healthy looking green leaf. That green leaf is probably the work of one of God's more amazing tiny insects.

Several species of beetles, flies and moths go through a larval stage in which the larvae burrow into leaves. These leaf-miners gain their nutrition from the leaf. But when fall comes, and the leaves lose their green chlorophyll, they stop making food for the larvae. So the larvae secrete a hormone that prevents the leaf from losing its chlorophyll and shutting down to die. This hormone enables a leaf to stay green and produce food for the larvae even long after the leaf would normally have turned color and dried up on the ground.

Leaf-miners present several problems for those who think that mindless evolution produced the variety of life that we see around us today. How did the larvae learn to make just the right hormone to preserve the leaf on which it depends? Are we to believe that these larvae kept evolving from something else until one of them figured out the chemistry of the leaf? And finally, are we to believe that this unlikely evolutionary event happened in several species?

Just as the sprouting of new leaves in the spring means that summer is near, so the cleverness of the leaf-miner should tell us that the hand of God created this insect. And we can reliably conclude that God made the entire creation.

Prayer: Dear Father, I thank you for the witness of your wonders. Amen.

Ref: Joachim Scheven, "Green Islands," *Origins*, June 1999

We Live in a Puff of Smoke

Matthew 24:30a
"And then shall appear the sign of the Son of man in heaven, and then shall all the tribes of the earth mourn..."

Classical evolutionary cosmology has the universe coming into being through a giant explosion of matter and energy called the Big Bang. The question that modern physics has struggled with for years asks whether there is enough mass in the universe to cause it to stop expanding at some point. Would the universe begin to collapse in on itself at some time in the future only to produce a second Big Bang?

In 1997, researchers announced that not only is the universe expanding, but also it is expanding faster now than it was shortly after its creation! The finding was based on the fact that very distant exploding stars are much dimmer than expected. After more than two years of further research and dozens of observations, scientists are beginning to agree that the universe's apparent expansion is increasing. Researchers' inability to explain why has led them to conclude that they don't have a theory about the universe that works. One researcher referred to this problem as "the biggest embarrassment" in modern physics.

So next time you hear a scientist speaking with certainty about the Big Bang, the age of the universe, or how it might end, you will know that, in truth, such certainty does not exist in the halls of science. For all they know, the universe could keep expanding until it dissipates like a puff of smoke. But we know from the Bible that the universe will end when our Creator returns to take those who trusted in Jesus Christ as their Lord and Savior home to a new and perfect Heaven and Earth.

Prayer: Dear Lord, help me to be prepared for your return. Amen.

Ref: James Glanz, "Calculating weight of emptiness is No. 1 cosmological mystery," *Star Tribune*, Sunday, December 5, 1999, p. A19

A Brain to See God's Handiwork

Acts 17:29

"Forasmuch then as we are the offspring of God, we ought not to think that the Godhead is like unto gold, or silver, or stone, graven by art and man's device."

One of the most difficult problems for those who believe that we are the product of millions of years of evolution is the human brain. The human brain has ten billion times 25,000 neural connections. If you work it out, this means that a miracle would have to take place if we evolved from an ape-like creature on the time scale that evolutionists propose: every generation would have had to have many thousands more neural connections than the last!

Even more astonishing is the fact that we are not born with a brain that is a blank slate. Several studies have shown that even infants as young as three months expect things to behave in certain ways. In one study, three-month-old infants showed surprise when researchers made an object they were looking at disappear. Another study showed that young infants expect inanimate things not to move by themselves. But they expect living things to move all by themselves. Such built-in "programming" in the brain enables us to progress more quickly as we learn about the world around us.

The problem of the complexity of the human brain, and the fact that it seems to come preprogrammed, ultimately caused Alfred Russel Wallace, an influential evolutionist, to switch to belief in a Creator. The more we learn about what God has created, the more likely we are to conclude that we are indeed the work of a wise Creator. He has even given us a brain capable of seeing that.

Prayer: I thank you, Dear Father, for the intelligence you have given me. Amen.

Ref: *Natural History*, 9/97, "The Nature of Learning," pgs. 42-45.

The Ancients Could Have Moved the Saturn V Rocket!

Job 38:36

"Who hath put wisdom in the inward parts? or who has given understanding to the heart?"

We look at Egypt's pyramids and wonder how the ancients built these great structures. Such questions are really about the source of man's intelligence and creativity. Were intelligence and creativity created within us by an ultimately intelligent and creative God, or are these a result of blind evolutionary forces?

In the ancient city that the Greeks called Heliopolis, the Romans converted an ancient Greek temple into a temple for Jupiter. The original temple must have amazed even the Roman builders. It is 2,500 feet long, making it one of the largest stone structures in the world. Twenty-six feet above the foundation of this structure are three of the largest stones ever moved by man. These limestone blocks measure 10 feet, by 13 feet, by 60 feet long. They weigh 1.2 million pounds each, three times the size of any stone used in the pyramids!

Not only were the stones raised 26 feet into the air, but also they were moved from a quarry that is three-quarters of a mile away. At the quarry is an even larger stone weighing two million pounds that would have crushed any logs used to roll it to the temple. The first time in recorded history that man ever moved anything so heavy was when NASA moved the giant Saturn V rocket to the launch pad on its now famous giant tracked vehicle.

The reason that these ancient people could have moved the Saturn V rocket is that intelligence comes from our Creator, not chance evolution.

Prayer: Dear Father, I thank you that you didn't use evolution to create me. Amen.

Ref: *Science Frontiers: Some Anomalies and Curiosities of Nature*, William Corliss, "Unbelievable Baalbek," pgs. 17-18

Research Challenges "Survival of the Fittest"

Jeremiah 31:2
"Thus saith the LORD: 'The people which were left of the sword found grace in the wilderness; even Israel, when I went to cause him to rest.'"

Charles Darwin popularized the phrase "survival of the fittest." Ever since, this idea that evolutionary change is driven by the necessity of survival has been essential to the idea of evolution. Evolutionists examine each feature of a plant or animal to try to determine why that feature evolved to help the creature survive. Researchers have modeled survival using two creatures with different features. Each feature gave the two creatures advantages over each other in differing situations. When conditions favor one of the creatures, it prospers, while the other suffers unless it evolves some advantage. While such simple models often appear in textbooks, such simplistic situations almost never happen in the real world.

Recently, researchers in the Netherlands constructed a more complex model of changing conditions based on the real world. They examined phytoplankton populations where 20 to 40 species of algae and diatoms can exist in a cubic centimeter of water. They used a computer to model the various needs of each species and the resources available to them. And when one species thrives, it sets up the conditions for other species to thrive as it uses up the resources it prefers. In the end, who's thriving among the 20 to 40 species varies over time, but there is no deadly competition for survival.

Our survival depends, not on the principles of survival of the fittest, but on the provision of our gracious Creator.

Prayer: I thank you, Dear Father, because my life depends on you. Amen.

Ref: *Science News*, "Algae need not be fittest to survive," 11/27/99, pgs., 340-341

Trilobite Eyes

Romans 1:20a
"For the invisible things of him from the creation of the world are clearly seen, being understood by the things that are made..."

A tiny parasitic insect has eyes and a lifestyle that are unique among living things. Actually, only the male members of the species *Xenos peckii* have eyes. The females spend their short lives inside the paper wasps they infest and don't have eyes. After a male hatches inside a wasp, it emerges from the wasp and uses its entire two-hour adult life span searching for a wasp infected with a female of its species.

The males' eyes are unlike the eyes of any other living thing. Each faceted eye has fifty lenses. And each of the bulging lenses has more than 100 photoreceptors. Researchers say that about 75 percent of the parasite's brain is devoted to processing the visual information collected by the eyes. This in itself shows design, since the males have only two hours to find a mate within another paper wasp.

While no other living creatures have a visual system like this, the structure of the eyes seems similar to the now extinct trilobite. While tiny and seemingly unimportant, this small parasite is a testimony to God's ingenious creativity. What's more, the fact that it has a unique visual system carefully designed for its unique way of living leaves no known creature for it have to evolved from.

Prayer: I praise you in wonder, Dear Father, for the way in which the creation bears witness to you as its Creator. Amen.

Ref: *Science News*, 12/4/99, p. 361, "Living insect with eyes like trilobites'."

How Do You Make Sweet?

Psalm 119:103
"How sweet are thy words unto my taste! yea, sweeter than honey to my mouth!"

How do you make something that is sweet? For that matter, how do you make any taste? I'm not talking about adding sugar to a recipe. Let's say that God provided you with all the atoms of the various elements you needed to make a molecule that tastes sweet. If you were to make such a molecule, you would need to know the construction of the sweet taste receptors on your tongue. You would need to know which molecular shapes would bind to those receptors. Then you would have to understand how all the different elements bind to one another and what molecular shapes they would have.

The complexity of this task was well illustrated when researchers at the University of Wisconsin-Madison released a computer model of a protein called brazein. Brazein is 2,000 times sweeter than sugar and found naturally in a West African fruit. Brazein's coiled molecular shape is responsible for its sweetness but if the molecule uncoils it loses all its sweetness. Its structure closely resembles scorpion poisons and proteins used by some plants for self-defense. The more we learn about how things work in the world around us, the clearer it becomes that everything that exists has been carefully designed to interact in very specific ways. We learn more about this Creator, though, not from molecules, but from His holy Word.

Prayer: Dear Father, grant me understanding as I study Your Word. Amen.

Ref: *Science News*, "Protein's shape may give extra-sugary shape," 6/20/98, p.389

Magnetic Monarchs

Psalm 89:12
"The north and the south, thou hast created them; Tabor and Hermon shall rejoice in thy name."

Scientists have always wondered whether monarch butterflies have built-in compasses to help them in their fantastic migrations. After wintering in Mexico, monarchs head to their northern breeding ranges in the eastern United States and Canada. Flying 1,000 miles a day and never having seen their breeding range, they find it every year. By the time they begin their migration south, they are the great-great-grandchildren of the individuals who left Mexico the previous spring. Yet, never having seen their wintering area, they find the same area in which their distant relatives wintered.

Researchers placed monarchs that were migrating south into a room shielded from the Earth's magnetic field. When the Earth's magnetic field was allowed to be normal in the room, released butterflies flew to the southwest, as they should. When researchers set the room to have no magnetic influence, monarchs flew in all directions. And when researchers generated a magnetic field in the room opposite that of the Earth's, the butterflies flew northeast. Researchers add that this demonstration of compasses within the monarchs still does not fully explain the wonders of monarch migration. Here at Creation Moments, we would say that it takes more faith to believe that monarchs and their migration habits evolved than to believe that these are the work of a wise and intelligent God Who also gave the Earth a magnetic field.

Prayer: Father, I am filled with wonder at all you have made. Thank You. Amen.

Ref: *Science News*, "Monarch butterflies use magnetic compasses," 11/27/99, p. 343

The Most Bitter Substances in the World

Proverbs 27:7
"The full soul loatheth an honeycomb, but to a hungry soul every bitter thing is sweet."

At least one-ninth of all 900 species of gourds produce chemicals known as cucurbiticins. The ten cucurbiticins so far identified are the most bitter substances known to human taste. In tests, subjects were able reliably to identify amounts as diluted as one part per billion! Cucurbiticins are also poisonous. Livestock have been known to die from eating gourds containing the chemical. But most plant eaters, including insects, know to stay away from the gourds that contain this chemical.

Despite the bitterness of cucurbiticins, a family of 1,500 species of beetles, including cucumber beetles, love to eat the chemical. They can detect the chemical in plants several yards away, or even further if the gourd is injured. When these beetles find a gourd with the chemical in it, researchers say they almost seem compulsive about eating it. As a test, researchers laced tiny grains of sand with cucurbiticins. The beetles actually ate the sand!

Much of the bitterness in our lives comes when we try to live life by our own rules, rather than the rules of our Creator. Think of the Bible as an owner's manual for your life on Earth. Not only does it tell us how God designed us to live, but it tells us of our Savior, Who came to rescue us from the consequences of our disobedience to God.

Prayer: Dear Father, I thank you because you sent Your Son to rescue me from the bitter results of sin in my life. Amen.

Ref: *Bombardier Beetles and Fever Trees*, William Agosta, pgs. 10-12

How Long Does it Take to Make a Fossil?

2 Samuel 22:47
"The LORD liveth; and blessed be my rock; and exalted be the God of the rock of my salvation!"

How long does it take to make a fossil? Walk through just about any museum and you will find fossils with explanations saying that they are "millions of years old." Many people find this claim plausible since the fossils are often from creatures very different from those we see today. And for many people, this is convincing evidence that the story of the evolution of life over millions of years is true. These fossils convince many that young earth creationism is wrong.

This line of evolutionary thinking would be shown to be completely invalid if it could be demonstrated that fossils don't need millions of years to form. While most fossils don't come with labels that allow you to work out their age, a recent fossil find off the coast of Victoria, Australia does. It is a ship's bell, firmly encased in solid rock. In former times, ship's bells carried the name of the ship on which they served. Found in about three feet of water, this fossil identifies itself as being from the sailing ship *Isabella Watson*, which sank off that coast in 1852. That means that this fossil, and the rock in which it is encased, is less than 150 years old!

This is but one of several examples of young fossils. Each of these young fossils demonstrates that it does not take millions of years for fossils to form. Every fossil ever found fits easily within the time line provided by young earth creationism. This also means that the Rock we should be studying is Jesus Christ, the Rock of our salvation.

> *Prayer: Lord, I hope on you for my salvation. Thank you for saving me. Amen.*

Ref: *Creation*, March-May 1998, p. 6.

Has the World's Oldest Shoe Store Been Found?

Exodus 3:5
"And he said, Draw not nigh hither: put off thy shoes from off thy feet, for the place whereon thou standest is holy ground.'"

While evolutionists portray early man as unskilled and primitive, the Bible portrays him as skilled and sophisticated, even by today's standards. Martin Luther, in describing Adam before the fall, said that we today would probably consider him to be a type of superman. The Bible's view of ancient man was supported when researchers found a collection of 18 shoes and sandals in the Arnold Research Cave in central Missouri. While some of the shoes were dated at about 1,000 years old, others have been radiocarbon-dated at many thousands of years old.

The shoes impressed researchers with their excellent design. Two of them are made of leather; the rest are made of plant fibers. They include both sandals and slip-ons. Those woven from plant fibers are equally sophisticated, whether they are only 1,000 years old or many thousands of years old. Left-and right-foot shoes and sandals are easily evident, and some even show toe impressions. Archaeologist Michael J. O'Brian of the University of Missouri and Columbia summed up the quality of the footwear, saying, "These people certainly knew what they were doing."

The shoes were so well designed that there was no need to improve the design over thousands of years. This supports the biblical view of man's history. Man's skill and intelligence were gifts of God from the beginning.

Prayer: Dear Father, I thank you for the skill and intelligence you have given me. Help me to use them for your service. Amen.

Ref: *Science News*, 7/4/98, "Ancient North Americans shoes step to fore";
Daily News, 7/9/98, "Ancient shoes show fashion is nothing new."

Could the Legend of the Phoenix Be Based in Fact?

Acts 27:12a
*"And because the haven was not commodious to winter in, the
more part advised to depart thence also, if by any means they
might attain to Phenice, and there to winter, which is a haven of
Crete..."*

The legend of the phoenix goes back thousands of years.
According to that legend, the phoenix bird lived for hundreds of years in
the Arabian Desert. Then, to renew itself, it burned itself up in its own
funeral pyre. It would then rise from its own ashes, renewed for several
hundred years.

Dr. Maurice Burton, a British naturalist, has suggested that this
legend may have some basis in fact. He points out that some birds like to
play with fire. The British rook is a bird that is a little larger than a crow.
When Dr. Burton hands a rook at his nature preserve an unlit match, the
rook holds the match so it can peck at it. Once the rook gets the match to
light, it quickly puts the burning match under its wing, appearing to want
to set itself on fire. When supplied with straw and a match, the rook will
set the straw on fire and then lie, wings outstretched, on the burning
straw until the fire goes out. It is possible that in ancient times people
saw this behavior and developed the legend of the phoenix to explain
what they saw.

Dr. Burton points out that this behavior is common among
intelligent birds, such as rooks and jays, who pick up discarded cigarettes
that are still burning and fly off with them. It is even likely that this
strange behavior is responsible for many house fires. While the legend of
the phoenix may be a misinterpretation of a real event, we can trust that
the Bible accurately records real events from creation to our salvation to
eternity.

**Prayer: Father, I thank you that I can trust your Word in all
that it teaches. Amen.**

Ref: *Toledo Blade*, 10/8/59, "Phoenix Rises Again In Bird Fire Addicts."

Where Is Your Hope?

Romans 15:4
"For whatsoever things were written afore time were written for our learning, that we through the patience and comfort of the scriptures might have hope."

Where do you find hope? Christians find their hope outside themselves in the sure promises of God in Jesus Christ. These promises are found throughout the Bible from Genesis to Revelation. But what happens to our hope when we decide to be selective about what we want to believe of the Bible's teachings and or when we add other beliefs that are not from the Bible?

One good example of what happens can be found in Jane Goodall's recently released autobiography, *Reason for Hope: A Spiritual Journey*. Goodall became famous for living with and studying chimpanzees in Tanzania. While her family doesn't appear to have had a strong Christian faith, Jane's grandmother is described as having a very deep faith and wishing the same for Jane. By her own accounting, by the time she was in her teens, Jane decided she would be selective about the Bible's teachings, believing what she wanted and discarding the rest. Then she started adding beliefs from non-Christian religions to her belief system. She admits that this left her with the problem of where to find hope.

In her autobiography, Goodall writes that she finally found four reasons for hope. They include, the human brain, the resilience of nature, the indomitable human spirit and the energy and enthusiasm of young people. But, we add, in a sinful world all of these things are more likely to lead to suffering and evil than to provide hope. Our only real hope is found in Jesus Christ and His victory over sin, death and the devil.

Prayer: Dear Father, I thank you for the hope you give me in Christ. Help me to share my hope with others. Amen.

Ref: *World*, 10/30/99, "Scientists need faith."

Colorful Salmon Communication

Romans 12:18
"If it be possible, as much as lieth in you, live peaceably with all men."

When salmon hatch, they remain in the rivers where they hatched for one to four years. There, they establish feeding territories and grow in preparation for the more challenging life at sea. This means that a river may contain newly hatched salmon in competition with salmon who have a several-year-old territory. The salmon have been created with a unique way of holding their territories without hurting each other.

Young salmon have darker ovals on their back and sides, as well as a thin stripe around each eye. As the young salmon challenge each other for territory, the salmon that senses it is not prevailing darkens these patches on its skin. When the fish on the losing end of the battle decides to abandon its challenge, these patches darken and the thin stripe around the eye becomes a thick, dark ring. This ends the challenge.

Researchers at the University of Glasgow in Scotland studied 40 challenges among young salmon. They found that when fish were forced into competition, most of the time the loser showed the coloring. In those instances where no darkening was shown, researchers concluded that the circumstances provided no real competition.

This amazing finding shows us that salmon have one more thing to teach us. They have been created with a system where no one gets hurt in their confrontations. Those of us who have been recreated in Christ are also sometimes forced into confrontations. But if we cultivate Christ's humility, we can settle the confrontation in a peaceable manner.

> ***Prayer: Dear Father, thank you for recreating me in Christ. Help me be humble so that I might live peaceably with all. Amen.***

Ref: *Science News*, 12/11/99, p. 375, "Fighting salmon fly dark flag to surrender."

This Serpent Bears Witness to God's Greatness

Psalm 145:15
"The eyes of all wait upon thee; and thou givest them meat in due season."

When you are a small animal whose diet could injure or even kill you, eating can be positively dangerous. That danger increases if you belong to a family of animals that normally takes minutes or even hours to swallow a victim.

An adult Texas thread snake is only about six or eight inches long, a little thicker than a strand of spaghetti and has a jaw unlike any other creature. It eats ants—ants that could seriously injure or kill it. As you may know, snakes generally take several minutes or even hours to swallow their food as they slowly stretch their jaws around their prey. But this method wouldn't work for the thread snake. By the time it was swallowed, the snake's lunch would have killed the snake. So the thread snake has a lower jaw, filled with teeth, and a hinge in the middle. As the lower jaws hinge in and out—several times a second, like a swinging door—ants disappear into the snake in the blink of an eye. Thus they are able to down dozens of ants a minute!

The Texas thread snake is more than a witness to God's wise design. Its unique jaw structure is essential to its survival. If evolution is true, the Texas thread snake would have become extinct during the millions of years it waited for evolution to find a solution to its problem. In addition, since there is no other vertebrate with a jaw like the Texas thread snake, from what could it have evolved? Unlike the serpent in the Garden of Eden, the thread snake truly bears witness to God's greatness.

Prayer: I thank you, Dear Father that you have provided for the needs of all your creatures in a way that witnesses to you. Amen.

Ref: *Science News*, 12/4/99, p. 361, "Weird jaws let tiny snake gulp fast."

Your "Little Angel" Isn't

Psalm 51:5
"Behold, I was shapen in iniquity, and in sin did my mother conceive me."

Modern psychologists and social engineers generally hold that children are born innocent. Any evil they do is learned from those around them. Based upon this notion, they advise that children should not be spanked because all that spanking teaches is violence. This modern teaching contradicts the biblical teaching that we are born in original sin, inherited from Adam. Belief in original sin is accepted in some form by most Christians, Jews and practically everyone who has ever raised children!

Now science has stepped into the debate. Psychologist Richard Tremblay of the Universite de Montreal studied 511 children, all under 18 months old. He found that 70 percent of the children grab things. Forty-six percent push others, 21 percent physically attack others, 23 percent fight, 27 percent bite and 24 percent kick. These evil behaviors were going on long before the children could have learned them from those around them. Psychologist Tremblay concluded from his study that the parents' real task is not to teach children to be themselves but rather to teach children to obey moral principals.

This research supports the biblical teaching of original sin, and by implication, supports the Bible's account of Adam and Eve. One theologian pointed out that this explains why parents must daily battle against self-idolatry, and must do this for their children until their children learn to do this for themselves. This study simply supports what the Bible has been saying for thousands of years.

> *Prayer: Dear Father, I praise you for you have given me salvation, which because of original sin, I could not earn myself. Amen.*

Ref: *Alberta Report*, 8/16/99

Simple to Complex or Complex to Simple?

Proverbs 12:5
"The thoughts of the righteous are right, but the counsels of the wicked are deceit."

A strange and beautiful family of creatures known as ammonoids nicely illustrate the illogic and lack of scientific reasoning used by defenders of evolution. While the chambered nautilus is the only species of ammonoid that exists today, the fossil record reveals hundreds of extinct species. Some, like the chambered nautilus, had relatively simple shells, while others had intricately detailed shells with ruffled, wavy walls.

Historically, evolutionists have always explained this variation by saying that, like all other living things, ammonoids evolved from simple to complex. Recently an ammonoid specialist developed a system that mathematically rates shell complexity, based on factors like the length and number of loops and spirals in the shell. He concluded that even though the relatively simple chambered nautilus survives, those with more complex shapes had a survival advantage in the past. He argues that the ammonoids evolved from complex to simple, thus proving evolution.

Simple to complex? Complex to simple? Both arguments cannot be used in science to support evolution. The fact that supporters of evolution illogically argue in both directions shows that evolution is not scientific. It is merely a philosophy that seeks to explain the creation without need for the Creator. But ultimately, evolution is unable to explain the creation without our Creator.

Prayer: Lord, deliver us from the deceitfulness of unbelievers. Amen.

Ref: *Toronto Star*, 11/21/99, p. F8, "Survival of the simplest."

God's Healing

Matthew 9:12
"But when Jesus heard that, He said unto them, 'They that be whole need not a physician, but they that are sick.'"

Most people know that diabetes is a disease that affects the body's ability to regulate the production of insulin. But few know that 60 percent of those with diabetes also suffer nerve damage. This nerve damage can result in the body's inability to control blood pressure as well as other problems. In some diabetics, branching nerve cells swell, cutting off normal cell communication.

Medical researchers have been testing a growth factor known as IGF-I to see if it will stop this damage. To their surprise, not only did IGF-I stop the nerve damage in diabetic rats, it also reversed the damage that had already been done! Tests are continuing to see if this treatment is safe for humans.

Do advances like this mean that God is becoming less important in the healing process, as many think? The truth is, all healing comes from God. So when you pray for healing, know that if it is God's will, God will heal you. In our usual experience, He uses modern medicine to do that, but if He wants, He will simply touch you with good health.

More accurately, modern medicine is a blessing from God for our good. Despite its accomplishments, modern medicine can never heal the root cause of our illnesses—sin. Only Jesus Christ can do that.

Prayer: Dear Father, I ask for good health, and I thank you for the spiritual health you have given me in Jesus Christ. Amen.

Ref: *Science News*, 12/11/99, p. 381, "Compound reverses diabetes damage."

The Myth of the Walking Whale

Psalm 148:7
"Praise the LORD from the earth, ye dragons, and all deeps..."

Whales pose some interesting problems for those who believe that all living things slowly evolved to their present forms over millions of years. According to evolution, sea creatures gradually adapted to life on land because they could make a better living there. That's where mammals supposedly evolved. Then, for some unknown reason, some of those mammals—Darwin said it was the bear—decided to return to the sea. This required the mammal ancestor of the whale to lose its legs, readapt to locomotion in water, develop new vision abilities, and move its breathing nostrils to behind its brain.

As unlikely as all of this sounds, some evolutionists make one more amazing claim. They claim that some whales still have the vestigial bones of their pelvises or legs embedded in their bodies. Some have even claimed that certain whales have been seen that still had vestigial legs growing out of their bodies. (No trace of these so-called "legs" can be found in any scientific literature.) The small bone that some whales have is not a vestigial pelvis. It is not even attached to the backbone or any part of the skeleton but is situated within the body as an anchor for some of the whale's organs. In short, any talk of whales with legs or vestigial walking structures is pure myth.

God created the whales, just as the Bible says. The great whales glorify God with their great majesty and power.

Prayer: With the whales, Dear Father, I glorify you for your work of creation and for my salvation in Jesus Christ. Amen.

Ref: *Creation*, 6-8/98, pp. 10-13, "The strange tale of the leg on the whale."

Plants Are More Complex than Evolutionists Thought

Job 42:2
"I know that thou canst do everything, and that no thought can be withholden from thee."

According to those who believe that all living things are a result of evolution, life evolved from simple to complex. Plants, which are simpler than animals, are believed to have evolved before animals. When we started learning about the genetic codes of living things, it was expected that plants would have simpler genetic codes than complex living things. This is turning out not to be the case.

Almost 300 researchers have recently completed their work to learn the exact DNA sequence of two of the five chromosomes of the wild mustard plant. They learned that a chromosome numbers two and four of the plant have almost 8,000 genes. By contrast, human chromosome number 22 has only 550 genes. To be sure, humans are more complex than the wild mustard plant. But, researchers say, the wild mustard is definitely more complex than animals such as worms and flies! Scientists point out that the reason for the unexpected complexity is that while worms and flies can only adapt to limited environmental conditions and remain worms and flies, the mustard plant can adapt to a great range of conditions and produce very different types of plants.

These findings strike two blows to evolution. Plants can be genetically more complex than some so-called "highly-evolved" animals. Second, this genetic complexity is wisely designed to give plants greater adaptability in a wide range of conditions. Our Creator designed the DNA code with purpose; He certainly did not let it develop by chance.

Prayer: Lord, I thank you because all of your loving purposes are fulfilled. Amen.

Ref: *Science News*, 12/18-25/99, p. 389, "Chromosomes show plants' secret complexity."

Batty Cologne

Psalm 141:2
"Let my prayer be set forth before thee as incense, and the lifting up of my hands as the evening sacrifice."

When a young man prepares himself to meet his young lady he slaps on a little cologne, hoping to please her. It turns out that the males of at least some species of bats do the same thing. Researchers have recently discovered how a small North American tropical bat prepares to attract and keep his harem of up to eight females. In late afternoon the bat begins a ritual that takes more than half an hour to complete. The bat starts by licking scent sacs in its wings. Then it gathers secretions including urine, from various parts of its body, depositing each secretion on its wing sacs. Despite what you might imagine, the resulting odor is sweet and spicy.

Once he is finished primping, the bat goes to his harem and hovers above them like a hummingbird. The bat can hover for up to 15 seconds. As he hovers, he flaps his wings hard enough to disturb nearby leaves and he chirps at each female. At the same time, his frantic flapping distributes his special scent. The females respond to the scent by chirping back at him. Males will also spread their scent around hoping to attract more females to their harems. Researchers think this behavior is more than just ritual. They point out that the complex mix of scents in urine can indicate whether a creature is in good health. For example female mice will not respond to males whose urine indicates that they are ill.

Many creatures communicate through scent. Scripture even speaks of our prayers as rising to God as sweet-smelling incense. But only the cleansing blood of Christ can sweeten us before God.

Prayer: Thank You, Lord, for taking away the stench of my sin. Amen.

Ref: *Science News*, 1/1/00, p. 7, "Male bats primp daily for odor display."

Are Rocks Like Vaults or Sponges?

1 Corinthians 2:7
"But we speak the wisdom of God in a mystery, even the hidden wisdom which God ordained before the world unto our glory..."

One of the common methods of finding the age of rocks, and ultimately the age of the Earth, measures the decay of radioactive uranium into lead—the more lead, the older the rock. The method, however, makes some very big assumptions. First, scientists assume there was no lead in the rock when it first formed, but this is unprovable. Then, they assume that rocks are like locked vaults and that no uranium, lead, or the in-between decay elements can ever enter or leave the rock.

The fact is, numerous scientific studies show all these assumptions to be wrong. Studies have shown that commonly dated rock material can have lead in it as it crystallizes. Other studies have shown that lead, uranium and the other elements important for dating can be removed from the rock by simple weathering or other conditions. In other words, while evolutionary scientists compare rocks to locked vaults, we are learning they are more like sponges. This explains why evolutionary dating methods have found, for example, one part of a rock to be 30 times older than another part of the same rock.

The next time you hear someone say that certain rocks or the fossils in them are millions or billions of years old, just remember, rocks are not like vaults. They are more like sponges that gain or lose elements, making accurate dating impossible. There is no good scientific evidence that the earth is any older than the few thousand years indicated by the Bible.

Prayer: I thank you, Lord, that I can trust you as the Rock of Ages. Amen.

Ref: *Impact #319* [ICR], 1/00.

Irish Indians?

Matthew 28:19
"Go ye therefore and teach all the nations, baptizing them in the name of the Father and of the Son and of the Holy Spirit..."

It has long been debated whether Christopher Columbus or Leif Eriksson was the first European explorer to discover America. But it's beginning to look as though this debate is beside the point. One surprising clue is found in inscriptions in West Virginia. These inscriptions are in a language called Ogam. Ogam is a written, alphabetical language in which the orientation of vertical and angled lines to a central horizontal line determines each character. The inscriptions were made sometime between the early sixth and late eighth centuries A.D., almost a thousand years before Columbus.

The form of Ogam writing used in the inscriptions found in West Virginia was developed by the ancient Irish Celts. How did ancient Celtic writing find its way into West Virginia only six or seven centuries after Christ? The answer may be in the inscriptions. Each inscription offers a Christian message! One reads, "The season of the blessed advent of the Savior, Lord Christ." It has been suggested that the writers were Irish monks who, in obedience to the Great Commission, set off to make disciples of the American Indians.

There is evidence that people in the Old World commonly had knowledge of the New World before the Dark Ages. These Ogam inscriptions may easily be a sign of Christian missionary work in the New World.

Prayer: I thank you, Lord, for those who brought your Word into my life. Amen.

Ref: *Science Frontiers*, William Corliss, p. 31, "Ogam Inscriptions in West Virginia?"

Australian Dinosaurs

Ezekiel 32:2b
"...and thou art as a whaler in the seas, and thou camest forth with thy rivers, and troubledest the waters with thy feet, and fouledst their rivers."

If the biblical account of history is true, then man and dinosaurs lived at the same time. Some believe that this accounts for the almost universal legends of dragons. In many parts of the world, these legends were passed orally from generation to generation. Unfortunately, it takes only a few hundred years for such legends to become distorted. America was already a country when Europeans got around to settling in Australia. There they found that the local Aborigines had oral accounts of monsters. The Aboriginal people insisted these were real flesh and blood creatures.

Aborigines in the northern and eastern parts of Australia tell of the *burrunjor*. Their descriptions matched that of the allosaurus. As late as 1961 in this part of Australia a tracker reported a bipedal reptile 25 feet long. Aborigines in central Australia tell of the *kulta*, which is described very much like a diplodocus or apatosaurus. Like these dinosaurs, *kulta* lived in swamps and ate plants. Cave drawings in northern Australia depict similar creatures. Aborigines refused to settle on Lake Galilee in western Queensland because a monster lived in the lake. This creature, which they called a *bunyip*, sounds like a plesiosaur.

The fact that the Aboriginal descriptions so accurately match dinosaurs known to the rest of the world show that these are recent memories—exactly what we would expect if the biblical history is accurate.

Prayer: Dear Father, I rejoice because I can trust your Word of salvation. Amen.

Ref: *Creation*, 12/98-2/99, p. 27, "Australia's aborigines . . . did they see dinosaurs?"

Poison-Eating Insects

Psalm 71:17
"O God, thou hast taught me from my youth: and hitherto have I declared thy wondrous works."

St. John's-wort or Klamath weed is not native to North America. Until it was brought from Europe in colonial times, North American insects had never seen the plant. St. John's-wort quickly spread because it was poisonous and thus not eaten by most insects. The plant manufactures a poison called hypericin that is activated by light. This poison is powerful enough to kill even livestock that might be foolish enough to eat it.

But there are several types of insects that not only eat St. John's-wort, but thrive on it. Some beetles, larvae of moths and butterflies and leaf miners munch happily away on the plant. No, they're not immune to the poison. Instead they have figured out how to prevent light from activating the poison. The leaf miners chew tunnels inside the leaves, staying out of direct sunlight. The moth and butterfly larvae use a similar approach. Some even tunnel inside the stem. A European beetle and its larvae also eat the plant. Adult European Beetles have a thick shell that shelters their bodies from the light. Their larvae eat St. John's-wort only at night, and then, before the sun comes up, they burrow into the ground.

If evolution were true, these insects would have had only a few centuries to identify how the poison works and evolve a counter to it. It clearly makes more scientific sense to conclude that God taught these creatures what they needed to know to eat St. John's-wort without harm.

Prayer: I glorify you, Lord, for your wondrous works and perfect teaching. Amen.

Ref: *Bombardier Beetles and Fever Trees*, William Agosta, pp. 12-15.

The Stinking Poison Bird

Acts 14:2
"But the unbelieving Jews stirred up the Gentiles and made their minds evil effected against the brethren."

The hooded pitohuis is an evil-smelling, blue jay-sized bird that is native to New Guinea. The bird has been known to science since 1827, but it held a secret that only began to be uncovered in 1989.

An American graduate student made the discovery while studying another local bird, the bird of paradise. Unfortunately, hooded pitohuis kept getting caught in his nets. As he released the unwanted pitohuis, they would claw and peck at him. After he released the birds, he licked the wounds they had caused. As a result, his mouth began to burn and finally became numb for several hours. This led the graduate student to send dead hooded pitohuis to the National Institutes of Health for further study. The poison the birds produce was finally identified as one of the most poisonous substances known. It is hundreds of times more poisonous than strychnine. The amount of poison in one bird can kill 500 mice. Most amazing is that only one other creature produces the same poison, the poison dart frog of Central America.

Those who believe in evolution maintain that the ability to produce the poison must have evolved twice. But it's not that simple. Not only do both bird and frog, half a world apart, have to evolved this unlikely ability twice, but they both had to evolve immunity to their own poison. Ultimately, evolution is a faith. It is a faith that is poisonous to Christianity because it makes death natural rather than a result of the first Adam's sin. And without the fact of the first Adam, there is no need for the Second Adam, Jesus Christ.

Prayer: Lord, I thank you for your victory over sin, death and the devil. Amen.

Ref: *Natural History*, 2/94, pp. 4-8.

Iguanas that Grow Larger and Smaller

Jeremiah 31:2
"Thus saith the LORD: 'The people which were left of the sword found grace in the wilderness; even Israel, when I went to cause him to rest.'"

What do you do when food resources are scarcer than normal for an extended period of time? People not getting enough nutrition lose weight and eventually muscle mass. It would really make more sense to proportionately shrink your body size. After all, children require fewer calories than adults. It appears that this is just what marine iguanas and perhaps some other reptiles actually do.

Researchers studying the marine iguanas of the Galapagos Islands found that sometimes individuals appeared to get smaller. For 18 years they simply blamed measurement error. Then one of the scientists studied the 18-year record of measurements and realized that something was really happening. During an El Niño weather pattern when the algal beds offered less food for the iguanas, individuals shrank in size. In addition, the largest males showed the greatest amount of shrinking. If connective tissue alone was shrinking, the iguanas would only shrink by ten percent. But large males were shrinking by as much as 20 percent during a two-year-long El Niño. This means that even their bones must shrink! More research showed that the more an individual shrank, the better its chance of survival.

Iguanas need only concern themselves with survival in this life. As human beings, we need to be concerned about the next life as well. We can thank God that He has provided eternal life for us through Jesus Christ.

Prayer: Lord, I will thank you forever for forgiveness and salvation. Amen.

Ref: *Science News*, 1/8/00, pp. 20-21, "Famine reveals incredible shrinking iguanas."

The "Uncle Dragon Country"

Isaiah 52:10
"The LORD has made bare His holy arm in the eyes of all the nations; and all the ends of the earth shall see the salvation of our God."

Historians have dismissed many allegedly ancient maps as modern forgeries because the maps show features that the ancients supposedly knew nothing about. Among the more credible ancient maps are 20 variations of ancient Chinese maps that may be as much as 4,000 years old. Some of these maps are held in museums around the world. What's more, the maps accurately record features all over the world!

One of the more well known maps is the circular Harris map. China is depicted in the center of the map, surrounded by the rest of the world. The map contains labels for many features around the world. Early Chinese mapmakers were fond of naming a land after a distinctive creature found there. For example, Australia, which is accurately depicted as being south of China, is called "Land of the Fire Rat people." This is thought to be a reference to the red kangaroo, which is common there. North America is described fairly accurately as being about 3,300 miles east to west and north to south. Ancient Chinese mapmakers were fascinated with the Columbia River system of the American Northwest because on a map it looks like the Chinese characters for the word "dragon." Because of this feature, the map calls North America "Uncle Dragon Country."

The claims that 4,000 years ago man could not have known this much about the world are based on evolution. As more becomes known, the facts of history are combining with the facts of science to disprove evolution.

Prayer: Use me, Dear Father, to help spread the Gospel to all lands. Amen.

Ref: *Across Before Columbus?*, pp. 279-282.

Hazor's Biblical History Confirmed

1 Kings 9:15
"And this is the reason for the levy which king Solomon raised: to build the house of the LORD and his own house and Millo and the wall of Jerusalem, and Hazor, and Megiddo, and Gezer."

Secular archaeologists usually assume that the historical statements found in Scripture are overstatements or simply incorrect. It was with this attitude that archaeologists approached the ruins of Hazor to see if they could tell whether Solomon actually did add to the wall of Hazor and build other structures there, as the Bible says. Hazor was an important city on the primary military and trade routes between Israel and Phoenicia, Syria, Mesopotamia and Egypt.

Hazor had been destroyed by Joshua when Israel took possession of the Promised Land. Over the following centuries the site began to be inhabited by Israelites. By Solomon's time, the city reached a size of about 10 acres and was home to between 800 and 1,000 people. Archaeologists found that sometime during the reign of Solomon, the city wall was expanded, doubling the size. Other buildings also were added at this time. Excavations show that at Solomon's time the city was prosperous. The city's six-chambered gate is typical of the gates Israel built at this time. Gates of the same design also exist at Megiddo and Gezer, two other cities named as being fortified by Solomon, according the same verse that mentions the fortification of Hazor!

Again the Bible has been vindicated as presenting accurate history. We can be equally sure that the Bible's account of creation is accurate because the Bible is God's Word.

Prayer: Dear Father, I thank you that your Word is trustworthy in all things, for it tells me of my salvation in Jesus Christ. Amen.

Ref: B*iblical Archaeology Review*, 3-4/99, "Solomon's City Rises from the Ashes".

Amazing Winter Moths

Job 42:2
"I know that thou canst do everything, and that no thought can be withholden from thee."

As fall comes to the temperate regions of North America, Europe and Asia, many birds head for warmer climates and insects enter suspended animation. While it might appear that nature is shutting down for the winter, amazing things are happening. About fifty species of moths are awakening from spending the *summer* in suspended animation. Winter is when these moths are active, despite the fact that they don't have any of the biological antifreeze that some other creatures do. The metabolic costs of such chemicals would hamper their way of life.

Winter moths can live as they do even though they freeze below 32 degrees because they emerge from the leaf clutter on the forest floor where temperatures almost never fall below 37 degrees. But when the air temperature rises above freezing, the moths emerge, shiver for several minutes to warm up, and look for food. Maple sap is among their favorite foods. One stomach full of this high-energy food provides enough energy for one of these moths to hibernate all winter. When active, they are able to increase their metabolism by over 8,000 times, using the same amount of energy that sustains them all winter in a mere 30 minutes. Summer moths shed heat from the thorax, while winter moths conserve their heat with an entirely different heat exchange system.

Winter moths illustrate that there are no limits to what God can do. So if you are ever tempted to wonder how God could have made the entire creation in six days, just remember His winter moths.

Prayer: I am comforted, Almighty Father, because you love me. Amen.

Ref: *Natural History*, 2/94, pp. 42-48, "Some Like it Cold."

Pie-Eyed Problems

Romans 8:22
"For we know that the whole creation groaneth and travaileth in pain together until now."

The sea holds some of God's most creatively designed living things. Just as on land, the variety of creatures in the sea was much greater in the past. One good example of this is an extinct sea-going reptile known as *Ophthalmosaurus*. A full-grown *Ophthalmosaurus* was about 12-feet-long and shaped like a dolphin. What made it unusual were its eyes. They were each about the size of a dinner plate!

Scientists estimate that *Ophthalmosaurus* could see its prey clearly even in the dim light that is available 1,600 feet below the surface. Based on its body structure, *Ophthalmosaurus* appears to have been a strong swimmer. Adding these facts together, scientists concluded that *Ophthalmosaurus* regularly swam deep in the sea to find prey or avoid predators. This led researchers to examine *Ophthalmosaurus' bones* more closely. When a diver rises to the surface too quickly, the nitrogen dissolved in his blood forms bubbles that can block circulation and destroy tissue. Scientists wanted to know if *Ophthalmosaurus* also suffered from this condition, known as the "bends." They searched for and found depressions in the joints typical of the bends.

While God's creation was initially perfect in every way, Adam's sin introduced imperfection and suffering. Even *Ophthalmosaurus*, deep in the sea, suffered because of human sin. Thankfully, God has given us a solution for human evil in His Son, Jesus Christ.

Prayer: Dear Father, I thank you for giving me your Son's victory over sin, death and the devil. In Jesus' Name. Amen.

Ref: *Discover*, 1/00, p.28, "My What Big Eyes You Have."

Law of the Jungle: Cooperation

Hebrews 13:6
"So that we may boldly say: 'The LORD is my helper; and I will not fear what man shall do unto me?'"

According to evolution, biological history is "read in tooth and claw." In this view, living things compete with one another. The fittest survive at the expense of the less fit. In a forest, this means that trees compete with one another for light, water and nutrients. Taller trees benefit from the light that they prevent shorter trees from receiving. But as it turns out the true law of the jungle is cooperation.

What goes on beneath the forest floor is essential to the forest itself. Tree roots gather water and nutrients from the soil. In addition, fungi live among the roots, feeding off the sap and other carbon compounds produced by the tree. In return, the fungi help make nutrients in the soil available to the tree. The picture of cooperation goes even further than this. Researchers have learned that the trees themselves cooperate with one another. This cooperation even exists between species. Researchers shaded some trees, leaving others in the sun. Tagging trees with different isotopes of carbon, scientists were surprised to find carbon compounds made by the sunbathed trees present in the shaded tree! The trees that were doing well were helping the trees that were not able to photosynthesize—even if they were a different species.

The true law of the jungle turns out to be cooperation. Rather than survival of the fittest, this cooperation between living things reveals a carefully designed creation made by a loving Creator.

Prayer: You, Lord are my helper. Help me to love others and help them as you have loved and helped me. Amen.

Ref: *Creation*, p.56, "Sylvan Symphony."

Britain's Dinosaurs

Genesis 8:19
"Every beast, every creeping thing, and every fowl, and whatever creepeth upon the earth, after their kinds, went forth out of the ark."

Many lines of evidence support the Bible's claim that Noah saved two of every kind of creature on the Ark, including dinosaurs. This means that humans and dinosaurs lived at the same time, rather than separated by millions of years, as evolutionists claim. Long before modern science discovered dinosaurs, many histories offered descriptions of these creatures.

Britain's history contains hundreds of stories about the large reptiles we now call dinosaurs. Even if all these accounts are not accurate, it is unlikely that so many similar stories were invented. According to ancient historical accounts, a large reptile killed and ate King Morvidus in about 336 B.C. In 1405, villagers near Sudbury drove what they called a large dragon into a swamp after the creature terrorized the area. It had killed and eaten a shepherd, then proceeded to eat the sheep. It was said that it had a huge body, a crested head with sharp teeth and a long tail. As recently as a hundred years ago, the older Welsh residents of Penllin in Glamorgan reported winged reptiles that lived in a colony in the nearby woods. They were described as brightly colored and reportedly stole the villagers' chickens.

It would appear that the reason most dinosaurs became extinct after the Flood has as much to do with human expansion as it does with a cooler, dryer Earth. And these widespread stories are just what we would expect if the biblical history is true.

Prayer: Dear Father, I praise you for your mercy that saved mankind twice. Amen.

Ref: *After the Flood*, Bill Cooper, pp. 131-135.

Potato Self-Defense

Psalm 91:3
"Surely he shall deliver thee from the snare of the fowler and from the noisome pestilence."

The Peruvian Andes gave the world the potato, which is one of the largest crops in the world today. The problem is, the modern potato is plagued by potato beetles and aphids, which also carry some viral diseases to the plant. Now a wild potato found in the Bolivian Andes is being studied because it seems to be quite resistant to these pests.

The Bolivian potato defends itself using tiny hairs on its leaves called trichomes. There are roughly the same number of tall and short trichomes on the leaves, and they work together to provide a complete defense system for the plant. The hairs are so closely spaced that even a tiny aphid cannot avoid them. The end of each hair has a sack filled with defense chemicals. The short trichomes release their chemicals only when disturbed. The longer trichomes continuously release their chemicals. When an aphid gets on a leaf, its legs cannot avoid coming in contact with the sticky fluid the short trichomes release. To make certain that the aphid collects enough of the sticky stuff, the plant also releases a chemical that agitates the aphid. This gooey liquid finally sets up into a hard mass, leaving the aphid to starve. It also appears that a scent emitted by one of the trichomes makes the leaves completely distasteful to potato beetles.

A potato may someday be developed that can save itself from the pestilence of aphids and beetles. But we cannot save ourselves from the pestilence of our sin. That's why God sent His Son, Jesus Christ, to die on the cross for our sins.

> ***Prayer: Lord, I thank you that I can depend on you for my salvation. Amen.***

Ref: *Bombardier Beetles and Fever Trees*, William Agosta, pp. 22-25.

Do Tattoos Have a Spiritual Meaning?

Leviticus 19:28
"You shall not make any cuttings in your flesh for the dead, nor print any marks upon you: I am the LORD."

The Old Testament explicitly forbids God's people from getting tattooed. It might be argued that believers in God have refused to get tattooed because they view the body as the Holy Spirit's temple. But new research is beginning to suggest that there is more to the issue.

Dr. William Cardasis, a Michigan criminal psychiatrist, sees a possible link between criminal behavior and tattoos. In his study of 55 patients at a maximum-security hospital, Dr. Cardasis has found statistical links between sociopathic behavior and the tendency to wear tattoos. He found that patients with tattoos were much more likely to have no regard for the rights of others, behave impulsively and lie and steal with no remorse. Another study of cadavers in New York City showed that the bodies of teen drug addicts had twice the number of tattoos compared to the general population. One tattoo artist told the press that he refuses to tattoo the face, neck or hands. He pointed out that some people consider tattoos in these places to be "serial killer territory." Dr. Cardasis adds that simply having a tattoo doesn't mean one is a criminal—that depends on what the tattoo means to the person wearing it.

Getting a tattoo is a permanent commitment to the symbol represented by the tattoo. Believers should have a permanent commitment only to Jesus Christ.

Prayer: Thank You, Lord, for making my body the temple of the Holy Spirit. Amen.

Ref: *The Toronto Star*, 12/26/99, p. J6, "What do tattoos say about the soul?"

Ancient Modern Steel

Isaiah 51:5b
"...the isles shall wait upon me, and on mine arm shall they trust."

The ages of human history are frequently divided into periods named after rocks and minerals. There was the Stone Age, the Bronze Age and the Iron Age. This evolutionary way of describing human history is based on the materials people used to make weapons and tools. On the other hand, the Bible describes man as developing his knowledge of metalworking within a few generations of his existence.

Researchers are currently studying an ancient steel factory in Turkmenistan. The 1,000-year-old steel factory produced the legendary Damascus steel, which was the strongest and most durable steel produced in ancient times. When the factory was operating, steel was heated to temperatures up to 2,500 degrees Fahrenheit. In the remains of the factory's crucible, researchers found traces of steel that told an amazing story. Some of the shards had traces of low carbon steel, while others had traces of high carbon steel. They concluded that the ancient steel makers were mixing the two together in a process called co-fusion. The result was strong, durable steel. What is amazing is that it was previously thought that co-fusion was discovered in the twentieth century!

While our knowledge may be growing today, we need to avoid false pride. The Bible teaches that the highest knowledge is the knowledge of God. His protection is better than any provided by the hardest steel.

Prayer: Dear Father, I trust your strength for my protection. Amen.

Ref: *Discover*, 1/00, pp. 20-21, "Medieval Metal Masters."

The Kangaroo Had to Have Hopped into Existence!

1 Chronicles 16:12
"Remember his marvelous works that he hath done, his wonders, and the judgments of his mouth..."

Those who believe that all living things are the product of millions of years of evolution tell us that living things develop features that aid their survival. They are quick to point out how some unusual feature of a plant or animal aids in its survival. That's why it's noteworthy that evolutionists cannot explain why kangaroos hop.

The kangaroo has several special designs built into it that make hopping its most efficient form of transportation. For example, every time it hits the ground, 70 percent of the kangaroo's downward energy is stored in its tendons to be released for the next hop. By contrast a running human only stores 20 percent of this energy. Also, kangaroos are designed in such a way that it is actually easier for them to breathe when hopping than when standing still. With each hop, the kangaroo's organs move within its body, pushing air out of its lungs. It uses its muscles only to inhale. Add all these features together and the faster a kangaroo goes, the less energy it uses for the same distance!

If the kangaroo didn't have these special designs, hopping wouldn't be an advantage to it. At the same time, if the kangaroo originally didn't hop, its special design features would be of no value. In other words, the kangaroo is a complete package of special designs that had to come into existence all at the same time, just as it left the hands of a wise Creator.

Prayer: Dear Father, I praise you because your marvelous works declare your glory. Help me to declare your glory. Amen.

Ref: *Creation*, 6-8/98, pp. 28-31, "Kangaroos: God's amazing craftsmanship."

Studies Link Violence with Self-Love

2 Timothy 3:2
"For men shall be lovers of their own selves, covetous, boasters, proud, blasphemers, disobedient to parents, unthankful, unholy..."

Popular psychology has for years declared that troubled young people, especially those who become violent, suffer from low self-esteem. But three studies released in the summer of 1999 conclude just the opposite: young people who become violent have too much self-esteem.

One of the studies, published by the American Psychological Association, observed 540 undergraduate students. After answering standard questions designed to measure self-esteem and narcissism, the students were put into different situations. They were given the opportunity to act aggressively against someone who had praised them, insulted them or did nothing to them. Researchers found that the most narcissistic students were the most likely to react violently. They also found that narcissists were especially aggressive against anyone who had offended them. Another study found narcissism is prevalent among prisoners convicted of rape, murder, assault, armed robbery and similar crimes. When their self-esteem was measured against the general population, it was found to be above average. The researchers involved in this study pointed out that the primary focus of prison rehabilitation is on building self-esteem. This, they concluded, is definitely the wrong approach, since such people already have an inflated view of themselves.

The Bible warns against narcissism. Many of the problems in our society today are undoubtedly the result of this condition.

Prayer: Father, it is enough for me that you loved me enough to save me. Amen.

Ref: *Toronto Star*, 12/99, p. J6, "Studies link violence and narcissism."

A Tinker Toy Fossil Embarrasses Scientists

Proverbs 12:20
"Deceit is in the heart of them that imagine evil, but to the counsellors of peace is joy."

In October of 1999 scientists announced the discovery of a new fossil at the National Geographic Society in Washington. Scientists from the Dinosaur Museum in Blanding, Utah, and the Institute of Vertebrate Paleontology and Paleoanthropology in Beijing, China, announced the discovery of a so-called "missing link." The feathered fossil was clearly a bird, but it had the tail of a small dinosaur. They named it *Archaeoraptor*.

The fossil from China was not excavated by scientists and its history was uncertain. It possessed the specialized shoulders and chest typical of birds, but it had the long tail identical to some small dinosaurs. *After* the announcement, scientists proceeded to study the fossil in detail. They found that a few bones that would have connected the tail to the bird body were missing. After more study, they concluded that they had been fooled. Someone had taken the fossilized tail of a dromeosaurid and made it look as if it had been part of the fossil bird. The scientists involved admitted they made a mistake by not examining the fossil before they announced the discovery of a "missing link." Like so many "missing links" before, this one was a hoax.

Every uncontested "missing link" that was ever "discovered" has suffered the same fate as *Archaeoraptor*. That these "discoveries" are announced before they are studied illustrates that evolution is nothing more than a faith. It is a deceitful faith, and the truth will only be found in the Bible.

> ***Prayer: Dear Father, let me never be deceived and led away from your truth. Amen.***

Ref: *Science News*, 1/15/00, p. 38, "All mixed up over birds and dinosaurs."

Evolution of Bats Gets Monkeyed Up

Psalm 33:6
"By the word of the LORD were the heavens made; and all the host of them by the breath of his mouth."

Bats defy evolutionary explanations. For their echolocation systems to evolve, bats had to develop simultaneously the ability to make high-pitched sounds, hear those sounds, and figure out what they mean. How did they eat before they evolved these abilities? Then there is the problem of evolving typical mammalian forearms into bat wings without crippling the creature in the process. Even evolutionists admit that the evolution of all these features even once is highly unlikely.

Bats are divided into two sub-orders. Smaller bats, like the free-tail bats of Mexico, are classified in the sub-order of *Microchiroptera*. Large bats like the fruit bat are classified into the sub-order of *Megachiroptera*. The brains of the larger bats have very different visual pathways than those in smaller bats. The visual pathways of the larger bats are more like those of primates! But no evolutionist would dare suggest that they evolved from primates. The second complication is that this means that both small and large bats could not have evolved from a common ancestor. It further means that all the unlikely features of bats had to have evolved at least twice, if evolution were true.

The Bible offers a simple explanation for the design of the bat and the differences between their brains. They were made by God using whatever designs He knew would be best for that creature's way of life, without regard for later classification systems.

Prayer: Lord, thank you for your Word that made life and gives life now. Amen.

Ref: *Creation*, 12/98-2/99, pp. 28-31, "Bats: Sophistication in Miniature."

156

Are We as Good as it Gets?

Proverbs 16:18
"Pride goeth before destruction, and a haughty spirit before a fall."

Are we, as human beings, as good as it gets? According to several influential evolutionary scientists, human beings are as good as it gets, and evolution has stopped for us. Some dispute this, saying that human evolution is still happening, but it's invisible. Although none of this sounds very much like science, such claims have fueled new debate among evolutionists.

These debates are puzzling, considering man's supposed evolutionary history. According to evolution, modern man evolved in a blink of evolutionary history from some ape-like creature. If evolution and its time scales were true, virtually every generation of our ancestors would have had to have evolved improvements over the last generation. Why would evolution have stopped so suddenly? Has modern medicine, which saves even the weak, stopped evolution?

Evolutions point out that mutations are decreasing among us. In the primitive past, women had children throughout their reproductive life. Older women give birth to children with more mutations. But today, women have their children when young, then stop reproducing.

The claim that we are as good as it gets reveals the pride of earthly thinking inherent in evolution. In truth, we can only be as good as we can get through the forgiveness of sins that comes by grace, through faith in Jesus Christ.

Prayer: Lord, I thank you for forgiveness and I look forward to the perfection I shall realize in heaven. Amen.

Ref: *Toronto Star*, 12/26/99, p. J7, "Are we the best our species will get?"

It Pays to Have a Second or Third Job

Genesis 1:25
"And God made . . . everything that creepeth on the earth after his kind and God saw that it was good."

Rove beetles are found in the rain forests of Costa Rica and have three different ways of making a living. Their largest prey are the blow flies that frequent animal dung. The beetles wait on the dung for a blowfly to land, and then stalk their prey with the skill of a cat. They can even capture their prey in the air if it tries to escape. But dung doesn't last long in the rain forest, because dung beetles quickly haul it off. So, as a second career, rove beetles also frequent corpses of dead animals. These are blowfly favorites, so rove beetles can make a good living on these, too. But, the rain forest has many creatures that clean up after a corpse, so they don't last long on the forest floor either. The rove beetle thus needs a third career.

That third career is carried out on leaves. Flies have little incentive to visit leaves. So the rove beetle has a special strategy to succeed in its third career. First, it positions itself on the leaf so that it looks like an innocent bird dropping, but smells like ripe and rotting fruit. This attracts fruit flies, which will approach the rove beetle unaware until it is snatched for lunch.

While the rove beetle and its way of life are not attractive to us, it is an important part of the cleanup crew in the rain forest. It, too, is part of God's design to make His complex Creation work.

Prayer: Dear Father, I praise you because everything you do is good. Amen.

Ref: *Natural History*, 11/94, pp. 18-23, "Masters of Deception."

Why There Was No October 5, 1582

Genesis 1:14
"And God said, 'Let there be lights in the firmament of the heavens to divide the day from the night; and let them be for signs and for seasons, and for days and years..."

Did you know that there was no October 5, 1582? There wasn't even an October 6 that year. Why? The story begins with the Earth's orbit. A year is exactly 365 days, 5 hours 48 minutes and 46 seconds long. This means that every year when the day of your birth arrives, the Earth is not exactly where it was in its orbit on the day you were born. Though the eleven-minute error is small, it does add up over hundreds of years.

The Julian calendar, with its leap day every four years helped correct this problem to some extent, but not completely. That's because the year is not quite 365 and one-quarter days long. Over the 1600 years of its existence, the minutes and seconds not reflected in the Julian calendar added up. So in 1582, Pope Gregory XIII announced refinements to the calendar by adding leap years. Moreover, every 400 years, a year ending in "00" received a leap day. This system keeps the calendar accurate to one day every 3,300 years. The old calendar had become inaccurate by ten days over its 1,600 years of use. To correct it, Pope Gregory declared that October 4, 1582, was to be followed by October 15!

The fact that man has always been making fairly accurate calendars reflects his intelligence, given to him from the beginning by an intelligent Creator.

Prayer: Father, help me use my time in your service and to your glory. Amen.

Ref: *Discover*, 2/00, "Fixing the Calendar."

Greek Pyramids

John 7:38

"He that believeth on me, as the scripture hath said, out of his belly shall flow rivers of living water."

When one hears the term "pyramid builders" one usually thinks of the Egyptians, Incas or the Aztecs, whose famous pyramids were built for religious purposes. You may not know that the Greeks also built pyramids. But as we might expect of the Greeks, their pyramids served a practical, everyday purpose.

Thirteen Greek pyramids have been identified; all are about 2,500 years old. The forty-foot-high pyramids are located in the hills around the cities of Theodosia and the Crimea. They are constructed of loose limestone rock through which the arid wind of the region could blow. As the cooling evening wind blew through the loose arrangement of rock, moisture in the air would condense on the rocks. The condensation would flow to the base of the pyramid where it was collected and piped to the city for use. A study of the clay pipes running from the pyramids led one archaeologist to calculate that each pyramid could produce a surprising 14,000 gallons of water per day!

These ingenious and practical pyramids reflect the truth that man has always been intelligent. Perhaps you didn't even know that water could be condensed from the atmosphere in usable amounts in this way. But the pyramids meant temporary survival in the dry climate. Permanent survival comes only from the living water that Christ gives through the forgiveness of sins. His water gives eternal life.

Prayer: Lord, help me share the living water you have given me with others. Amen.

Ref: *Science Frontiers*, William Corliss, p. 17, "Ancient Greek Pyramids?"

Could Creation and Evolution Be Telling the Same Story?

Genesis 1:1
"In the beginning God created the heaven and the earth."

People who believe that God used evolution to bring about all living things claim that the Bible and evolution are telling the same story using different words. They say that everything in both stories happens in the same order. But is this really true?

In the Bible, the Earth was made before the sun, while evolution insists that the sun existed before the Earth. Genesis tells us that sea creatures, which would include whales, were made a day before land animals. Evolution says that whales evolved from preexisting land animals. For that matter, the Bible says that land plants were made before life in the ocean, while evolution claims that life, including plants, started in the ocean. Genesis tells us that fruit trees were the first living things created. Evolution claims that fruit trees evolved fairly recently in geological history. The Bible teaches that birds were created before land animals. Evolution says that birds evolved from earlier land animals.

We could go on with still more contradictions between Genesis and evolution. It should be clear that they are not telling the same story. And there is one more, and most important contradiction between the two stories. While the Bible teaches that death is a result of our sin, evolution says that death is natural and was part of the evolutionary process. Evolution offers us no eternal hope. The Bible teaches that Jesus' death on the cross does offer us hope through salvation and eternal life.

Prayer: Lord, keep me in your Word so I may never be misled into false belief. Amen.

Ref.: *Back to Genesis [ICR]*, 6/99, p. d,"Could Evolution and Creation Be Telling the Same Story in Different Ways?"

Was Jesus Sending Us Today a Special Message?

Mark 8:24
"And he looked up, and said, 'I see men as trees, walking.'"

When Jesus healed people they were usually immediately and completely healed. In several instances, He wasn't even in the presence of the person He healed. Then we come to a curious instance of a healing in the Gospel of Mark. Mark 8:22-25 tells us that Jesus was asked to heal a blind man in Bethsaida. After spitting on His hands and touching the man's eyes, He asked the blind man if he could see anything. The man announced that people looked like trees. After Jesus touched his eyes again the man could see clearly.

Why couldn't Jesus, Who created everything in six days, Who could raise the dead, heal this man instantly and completely? Of course, He could have done that. But he seemingly chose not to. Perhaps He wanted to send a special message to people today. First-century medicine knew of no way to restore sight to those born blind, but modern medicine can sometimes restore the sight of those born blind. On receiving their sight, such people usually suffer from a condition known as agnosia. They can see, but their brains have not yet developed the connections necessary to interpret what they are seeing. Such people often say, when seeing for the first time, that people appear upside down and look like trees. Over time, the connections form between perception and reality.

So the healing of the man born blind was really two miracles. Perhaps Jesus wanted those of us who live in a time when such knowledge is available to recognize that these miracle stories are not just simplistic stories. They are medically accurate.

Prayer: Dear Father, I glorify you for preserving Your Word, which tells me of salvation in Jesus Christ. Amen.

Ref.: *Creation*, 9-11/99, pp. 54-55, "Walking Trees."

The Faint Sun Paradox

Genesis 1:16
"And God made two great lights; the greater light to rule the day, and the lesser light to rule the night: he made the stars also."

Let's assume for a moment that the evolutionary explanation for the origin of the Earth and sun is correct. The sun formed about 4.6 billion years ago. Let's imagine that evolutionists are right in saying that life originated on Earth about 3.5 billion years ago. This scenario creates many scientific problems, but today we want to focus on one problem that is seldom mentioned.

One implication of these assumptions is that the sun today is about 40 percent brighter than it was 4.6 billion years ago. This means that 3.5 billion years ago, the sun was bathing the Earth in considerably less energy than we receive today. Under those conditions life as we know it today could not have existed on Earth. Yet we are told it was under these conditions that life supposedly evolved. Evolutionists are aware of this problem. In an attempt to solve the problem they theorize that the early Earth had more greenhouse gases which would have held the sun's heat more efficiently and produced the same temperature range that life enjoys today. They then theorize that the mix of these gases gradually changed to the atmosphere we have today, perfectly matching the increase in the sun's energy output to Earth. This explanation is so unlikely that one scientific paper actually referred to it as the "Goldilocks syndrome!"

There is a much less fantastic solution to this problem. The Earth and sun are not billions of years old and life was created relatively recently to live under the conditions we enjoy today.

Prayer: With all that you have made, Dear Father, I glorify you. Amen.

Ref.: *Impact #300 [ICR]*, 6/98.

Chinese Indians?

Acts 21:2
"And finding a ship sailing over unto Phenicia, we went aboard and set forth.."

Today we look back on the European migration to North America as one of the results of a more enlightened age. When we do this, we forget that the American Indians migrated to North and South America from Asia and built their own culture long before the Europeans. If what a Texas Christian University linguist says is true, the Olmec Indians of the American Southwest and Central America may not have been Indians at all.

Linguist Mike Xu has spent several years studying 3,000-year-old Olmec jade, stone and pottery relics. The Olmec civilization appeared abruptly, as if from out of nowhere, about 1200 B.C. As he examined the hundreds of symbols on the relics it struck him that the Olmec symbols look very much like the Chinese writing of the same period. Olmec art is also very much like the Chinese art of the same period. He added that Olmec religious practices were very similar to Chinese religious practices of the time. For example both cultures put jade beads in the mouths of the dead to ward off evil. Xu concludes, "The similarities are just too striking to be a coincidence."

An entire civilization begun, and perhaps supported, by ships sailing from China implies a sophisticated transpacific shipping system. Perhaps we need to revise our evolutionary system for evaluating ancient history and give credit to people who were just as curious and resourceful as we are today.

Prayer: Lord, help me use the abilities you have given me to glorify you. Amen.

Ref.: *Discover*, 2/00, p. 20, "Chinatown 1000 B.C."

New Problems for Origin of Life Theories

Psalm 36:9
"For with thee is the fountain of life; in thy light we shall see light."

The usual theories for the origin of life have the first genetic material assembling itself by chance in a hot setting. Some have suggested that this may have happened in an undersea thermal vent or the side of a volcano. Trying to explain the origin of life without God suffers from many scientific problems. New research has now uncovered yet another problem.

Researchers at the University of California at San Diego examined how stable the chemical bases of genetic material are under various conditions. Origin of life theories must assume that wherever life began, these chemical bases had to build up to a concentration to make it likely that they would find enough of each other to make meaningful genetic material. This, they theorize, would have taken hundreds of thousands or millions of years. Researchers discovered that heat breaks down the four bases that make up genetic material. This effectively rules out thermal vents as a source for the first life. In a temperature at the boiling point of water one of the bases lasts only 19 days. None of them lasts long enough to build up enough concentration for life to start. Even at 75 degrees none of these chemical bases lasts long enough to be geologically important to evolution. Only freezing conditions allow the bases to last long enough.

All attempts to explain the origin of life without God's direct action and design are doomed to failure. Perhaps this is why evolutionists now suppose that life began in outer space!

Prayer: Dear Father, I rejoice in you as the source of life here and forever. Amen.

Ref.: *Proceedings of the National Academy of Sciences*, 7/98, pp. 7933-7938.

The Real Sin at the Tower of Babel

Genesis 11:6
"And the LORD said, Behold, the people is one and they all have one language, and this they being to do; and now nothing will be restrained from them, which they have imagined to do.'"

A misunderstanding of some of the history recorded in Genesis has led some people to conclude that we are repeating the sins committed at the Tower of Babel. After all, we build huge cities and incredibly high skyscrapers. English is now virtually a universal language and the Internet has led to the international pooling of knowledge.

The truth is that a universal language was not the reason God confused the language at Babel, although it contributed to the problem. Nor was creating a large tower the problem, although it too was a symptom of the people's pride. Pride certainly was a problem, and it led to the core sin committed at Babel. The core sin the people committed was defying God's command to subdue or inhabit the whole Earth. In failing to spread over the globe, they were only inhabiting a small part of it. God's command that people spread over the entire Earth was for our protection. He knew that if we were all one people that dangerous, dictator-led government could not be kept in check. God also knew that if one nation was infected with dangerous false philosophies there would be no other people who knew the truth.

In confusing their language at the Tower of Babel, God protected us from all of this. While we may not always understand the reason for God's commands, He always has our well-being in mind. That's because He is our loving Creator.

Prayer: Dear Father, I give you thanks and praise because you always have my well being in mind. In Jesus' name. Amen.

Famous Scientists Reject Darwinian Evolution

Proverbs 10:14
"Wise men lay up knowledge: but the mouth of the foolish is near destruction."

Modern evolutionists often give us the impression that evolution was accepted by the scientific community of Darwin's day. But the scientists of Darwin's day didn't hear his theory, slap themselves on the forehead, and say, "Of course!" In fact, some of the most influential scientists of the day were hostile to Darwinian evolution.

Take Adam Sedgwick for example. Although he taught field geology to Charles Darwin, Sedgwick flatly rejected Darwin's theory. After reading Darwin's book, he wrote to Darwin that parts of it were completely false and filled him with sorrow. Astronomer Sir John Herschel also rejected Darwin's theory. A fellow of the prestigious Royal Society, Herschel called Darwinian evolution, "the law of higgledy-piggledy." Evolution claims that life developed spontaneously by time and chance, but Louis Pasteur was motivated to prove that life only comes from life. Science philosopher William Whewell would not allow Darwin's book into the Cambridge University Library. James Clerk Maxwell strongly opposed Darwinian evolution.

Ironically, it was the churchmen of Darwin's time who embraced Darwinian evolution. And when those who grew up in evolutionist churches got their science degrees, all they knew was that "everyone" had always supported evolution. This illustrates how a church's lack of faithfulness can change the worldview of an entire culture for the worse.

Prayer: Lord, protect and preserve me from ever being led into false belief. Amen.

REF.: Creation, 9-11/99, pp. 26-27, "Holy War?"

167

Honeybees Tracked with Radar

Numbers 13:27

"Then they told him, and said, 'We came unto the land whither thou sentest us, and surely it floweth with milk and honey; and this is the fruit of it.'"

Honey has always been a popular food. For thousands of years it was a symbol of prosperity. Perhaps honey's continuing appeal is one of the reasons there is so much research on honeybees. Now, a research team has released findings that explain how honeybees learn how to navigate. Another team has released information showing how bees know how far they have traveled.

Before a honeybee can go out looking for food for the first time, it needs at least one training flight. Researchers cleverly attached tiny wires to bees to keep track of the bees' flight paths. They used radar tracking to make maps of the honeybees' flight paths. They found that after the trainee hovers in front of the hive, it makes a straight line away from the hive for a short distance, then returns. As training progresses, the trainee flies faster and further, demonstrating progressive learning.

The Israelites had to endure 40 years in the wilderness because they did not faithfully trust God to do what He said. As believers wander through the wilderness that is this life, we need to learn from Israel, and trust God's Word, even if we can't understand how God can keep His promises. His Word brings the message of salvation in Jesus Christ, which is sweeter than honey. Believing His promises will bring us to the promised land of heaven.

> *Prayer: Dear Father, I thank you for your Word and the salvation it has brought me through Jesus Christ. Amen.*

REF.: Science News, 2/5/00, p. 87, "Bees log flight distances, train with maps."

Does the Bible Provide Unnecessary, Impossible Details?

Genesis 35:17
"And it came to pass, when she was in hard labor, that the midwife said unto her, 'Fear not; thou shall have this son also.'"

Sometimes the Bible seems to offer unnecessary or even impossible details as it recounts a historical event. Take for example the Genesis account of Rachel's death while giving birth to Benjamin.

This account offers a very curious detail. When Rachel's labor is at its peak—with the baby not yet fully born—Rachel's midwife makes a seemingly impossible statement. She announces to Rachel that Rachel should take comfort in the fact that she is having another boy. In a normal birth the head emerges first, so it's impossible to tell whether the child is a boy or a girl until it is fully born. This detail might be enough for some to dismiss the account as nothing more than a fanciful legend. But a London physician has suggested that this account describes a breech birth. He suggested that Benjamin was born feet first, thus allowing the midwife to identify Benjamin as a boy as the largest part of his body was still emerging from the birth canal. This also explains why he was identified as a boy during the peak of labor. The physician further explains that in Rachel's day, breech births commonly resulted in the death of the mother.

Knowledge is lost when we question the accuracy of the Bible's account of history. Details in the Bible that seem unnecessary have often proven to be correct. Details in the Bible that seems impossible are only so because they are beyond our understanding.

Prayer: Lord, grant me understanding and insight into Your Word. Amen.

REF.: *Bible Review*, 2/98, p. 18, "Did Rachel Have a Breech Birth?"

Those Noisy Ants

Ephesians 4:29
"Let no corrupt communication proceed out of your mouth, but that which is good to the use of edifying that it may minister grace unto the hearers."

Scientists have long known that some ants communicate with one another with vibrations. They also know that these ants pick up vibrations through ears that are located in their knees. Ants in four subfamilies communicate using these vibrations. These ants tend to build nests with wood or dried pulp, which tend to carry the vibrations. Disturb a carpenter ant nest and the vibrations will begin. An individual ant will send out as many as seven vibrations in 50-millisecond intervals. Other ants in the nest will hear these vibrations as they travel through the material that makes up the nest.

New research has raised interest among researchers about whether some ants can hear airborne sound. To their surprise, ant researchers discovered that some ants communicate through audible squeaks. If you hold a red desert ant up to your ear, you may hear tiny squeaks. The squeaks are produced by the last segment of the ants body. The ant rasps two parts of the segment together in a principle similar to that used by crickets. Among the things communicated by squeaks are that dinner is ready, or a good nest site has been found. Squeaks are also used to call for help after a cave-in.

Communication is a wonderful gift of God, but the ability to communicate is not what makes us human. Rather, we are special in the creation because we were created in God's image and have been redeemed by His Son, Jesus Christ.

> *Prayer: Lord, I thank you that you have communicated your love to me. Amen.*

REF.: Science News, 2/5/00, pp. 92-94, "When Ants Squeak."

Air-Conditioned Dinosaurs

Job 41:20
"Out of his nostrils goeth smoke, as out of a seething pot or caldron."

Warm-blooded creatures need to be able to vent excess heat or they begin to suffer heatstroke. One of the ways in which man and warm-blooded animals do this is through our nasal cavities. As we breathe, the air passes through sheets of mucous membranes that are designed to increase the surface area over which the air passes. These membranes have a rich blood supply. When too much heat builds up, the membranes vent excess heat into the passing air, thus protecting the brain. The ingenious design of this system is an argument for an all-wise Creator.

Far from being "primitive," even large dinosaurs seem to have had this cooling system. Moreover, there is increasing evidence that many dinosaurs were warm-blooded. Researchers recently examined a *Triceratops* skull with a CT scanner and found that the nasal passages of the *Triceratops* occupied half of the skull. Skulls of smaller species of dinosaurs showed that they had disproportionately smaller cooling passages. It is believed that if the dinosaurs were warm blooded then the very large dinosaurs would indeed have required very large cooling passages.

There is some good evidence that the leviathan described in the book of Job was a large dinosaur. Large dinosaurs would need nasal cavities that transferred heat from their blood to the air they exhaled. Perhaps when Job describes smoke coming from leviathan's nostrils, he is describing heated, moisture-laden air hitting the colder air of the atmosphere as leviathan exhaled.

Prayer: I praise You, Lord, for I am fearfully and wonderfully made. Amen.

REF.: *Discover*, 2/00. p. 13, "Paleo-Pinocchios."

Carbon Dating Supports Noah's Flood

Genesis 8:2
"The fountains also of the deep and the windows of heaven were stopped, and the rain from heaven was restrained."

Did you know that radiocarbon dating may offer supporting evidence for Noah's Flood? Many think that carbon dating provides evidence that the Earth is millions of years old, but this is not true. Radiocarbon dating can only reliably date things that are less than 3,000 years old.

Let's say that you are a frog in a pond. High above your head, cosmic rays are striking nitrogen-14 atoms and changing them into radioactive carbon-14 atoms. That carbon eventually finds its way into your food and becomes part of you. Throughout your life it continues to accumulate in you. When you die, it stops accumulating. Being unstable, carbon-14 decays over time. Researchers measure the carbon-14 that remains in something that was once alive to determine how old it is. To do so, however, they must assume that the Earth is over 50,000 years old. That's how long it would take for the amount of carbon-14 being produced to equal the rate of carbon-14 decay. But we now know that there is more carbon-14 being produced than is decaying, meaning that the Earth must be *less* than 30,000 years old! In addition, the processes at work during Noah's Flood would have lowered the availability of carbon-14 to living things, which would also help account for the shortage of carbon-14.

In short, carbon dating cannot possibly tell us anything contrary to the Bible, and seems to actually support the Bible's account of Earth's history.

Prayer: Lord, you are my Rock of Ages now and into eternity. Thank you. Amen.

REF.: Back to Genesis [ICR], 7/98, p. d, "Doesn't Carbon Dating Prove the Earth Is Old?"

Egyptian Discoveries Enlighten, Support the Bible

Exodus 3:7
"And the LORD said: 'I have surely seen the affliction of my people which are in Egypt, and have heard their cry by reason of their taskmasters, for I know their sorrows.'"

The Bible tells us that at the beginning of the Israelites 430-year stay in Egypt, they were well respected for Joseph's sake. Later, when the Egyptian leadership forgot how Joseph saved Egypt, it began making their lives difficult.

One of the most enlightening archaeological finds from the period of Israel's captivity is a settlement called The Village of the Place of Truth. Excavators at the city have found homes, letters, notes, court records, work diaries and wills dating from this period. The village was home to those who built the royal tombs. While there is no evidence that the Israelites ever lived in the village, what has been found has revealed much about life in Egypt when Israel was there.

We learn that Egypt had a tradition in which every citizen put in at least one year of service to the government. Those out of favor could be forced into years or even a lifetime of forced labor. It was considered an honor to work on the royal tombs, however, and it seems unlikely that the Israelites had much to do with them. The Bible tells us that the Israelites did, however, make bricks. That could include the very bricks that built the Village of the Place of Truth. Finally we learn that pigs were a staple among such workers, which certainly would be repugnant to the Israelites.

Archaeologists have concluded that the Biblical record of Israel's stay in Egypt has too many accurate details about Egypt to be nothing more than a legend.

Prayer: I thank you, Lord, for knowing my sorrows and bringing me salvation. Amen.

REF.: Biblical Archaeology Review 1-2/99, pp. 36-45, "Pharaoh's Workers."

Did the Ice Man Use Acupuncture?

Jeremiah 17:14
"Heal me, O LORD, and I shall be healed; save me, and I shall be saved, for thou art my praise."

You may remember the "ice man" discovered in a glacier in the Alps in 1991. His frozen body had been preserved in the ice for thousands of years. Even his hair, beard and skin were largely intact. While there is some debate about when he lived, he is easily dated, in the Biblical framework, to within several centuries after the Flood.

Researchers at the University of Graz in Austria have been studying his body in hope of learning more about life in the distant past. They found that the iceman has 15 groups of tattooed lines on his body. They could find no reasonable explanations for the tattooed lines until they compared the tattoos to the traditional acupuncture points. The iceman's tattoos correspond to the acupuncture points used to treat stomach upset and backache. X-rays of the iceman's body showed that though he was only middle-aged, he did indeed have arthritis in his hips, knees, ankles and spine that would produce backache. Further study showed that he had a high level of parasites called whipworms in his intestines! The iceman lived 2,000 years before any known use of acupuncture by the Chinese. Could it be that acupuncture is ancient knowledge that came to use today by way of Noah himself?

Any medical treatment that works is a blessing in this life. However, only the healing provided by the Gospel of Jesus Christ can bring true spiritual health now and in eternity.

> ***Prayer: I praise you, Dear Lord, for the healing you have given me. Amen.***

REF.: *Discover*, 2/00, p. 61, "The Ice Man Healeth."

These Parrots Aren't Just Parroting

Psalm 94:10
"He that chastiseth the heather shall not he not correct? He that teacheth man knowledge shall not he know?"

Researchers at Purdue University are demonstrating that parrots are capable of intelligent communication. Carefully designed experiments have convinced even skeptics that the two parrots under study are not just giving conditioned responses. Their results are challenging accepted scientific knowledge about animal intelligence and the evolutionary claim that intelligence is one of the traits that separate us from animals.

The parrots' names are Griffin and Alex. Researchers can show Alex two triangles, one red and one green. Then, they ask what's different about them. Alex will answer, "Color." When asked what's similar about them, he will answer, "Shape." If there is nothing similar, Alex even understands the concept of nothing. Asked what's similar about the shapes, he will answer, "None."

Researchers say that about five percent of the time someone will mistakenly tell Alex that he has given a wrong answer when he has not. Alex, however, insists that his answer is right until the questioner discovers the mistake. Both parrots can identify a wide range of objects. Both parrots are even picking up words they have not been systematically taught and are using them correctly. Once when Griffin was having trouble pronouncing a new word Alex turned to him and said, "Speak clearly."

Evolutionists are now having to rethink the idea that intelligence separates us from animals. The fact that we were created in God's image to have a relationship with Him is what really separates us from the animals.

Prayer: I rejoice, Dear Lord, that I can have a relationship with you. Amen.

REF.: *Discover*, 1/00, pp. "Polly Want a PhD?"

Fossilized Fence Roadblocks Evolutionary Dating

Luke 19:40
"But He answered and said to them, 'I tell you that if these should hold their peace, the stones would immediately cry out.'"

When you see a fossil, or hear the word, you should not automatically think millions of years. In an attempt to maintain the illusion that fossils automatically must be millions of years old, evolutionists dismiss the examples of young fossils as freaks of nature. However, there are many examples of recent fossils.

It was low tide when a circular rock was discovered on a western Australian beach. A little over two feet in diameter, the wheel shaped rock weighed 165 pounds. It was hard and solid, even ringing like a bell when struck. Examination proved that the rock was a dense high calcium sandstone. Even before it was cut open for study, the rock seemed unusual. Once it was cut open it became clear that this was fossilized fence wire. Further analysis identified the wire as "Number 8" fence wire that was used at a nearby sheep station between 1920 and 1970. It was standard practice that after ten years, old fence wires were replaced. The old wire was coiled and often thrown of into the sea. This fossil could have been only about 30 years old and not more than 80 years old!

The existence of fossils does not disprove the Bible's account of history, which dates the Earth at only about 6,000 years old. If fossils can form in a relatively few number of years, 6,000 years are more than enough time to form all the millions of fossils discovered.

Prayer: Lord, protect me from being misled by those who doubt your Word. Amen.

REF.: *Creation*, 6-8/98, p. 6, "Fascinating Fossil Fence-Wire."

Scientists Recognize that We Have a Designer Earth

Isaiah 45:12
"I have made the earth, and created man upon it: even my hands, have stretched out the heavens, and all their host have I commanded."

Evolutionists often claim that evolution can produce living things that look as if they are designed. This is their way of answering arguments that things that look designed don't need a Designer. In January of 2000, however, a paleontologist and an astronomer teamed up to publish a book that says that conditions on Earth are unique—there is probably no other life like us in the universe. (We do need to keep in mind, however, that they write from an evolutionary perspective.)

Scientists Peter D. Ward and Donald C. Brownlee concluded that conditions could exist elsewhere in the universe that could support microbes. But, there may be nowhere else in the universe that life above that level might exist. First, we can rule out life near the centers of galaxies because lethal levels of several types of radiation exist there. The scientists conclude that the Earth is perfectly placed for intelligent life. Our moon is just the right size to control our climate, tilt and tides. The planet Jupiter acts as a giant shield to protect us from meteors and asteroids, and it, like the Earth, has an orbit that does not threaten other planets. Our sun is a heavy-element star with a rare elliptical orbit. And earth's atmosphere has a carefully balanced mix of elements with just enough carbon to support life.

Biological evolution cannot explain these delicate and unusual designs but the designs can easily be accounted for if they came from the hand of an all-wise, all-knowing Creator.

Prayer: In wonder and awe, I praise you, Dear Father, for your creation. Amen.

REF.: *World*, 2/19/00, p. 8 "Custom-made Earth: OK, but who made it?"

The Romans Beat Columbus to the New World

Psalm 104:25-26a

"So is the great and wide sea, wherein are creeping thing innumberable, both small and great beasts. There go the ships..."

Several evidences strongly suggest contact between Europe and the New World before Columbus. Some evidence even suggests that knowledge of the New World was common before the Dark Ages. Now mainstream science has concluded that the Romans made it nearly as far as Mexico City by the second century A.D.

Anthropologist Roman Hristov studied a black terracotta head that was uncovered near Mexico City in 1933. It had been unearthed by professional archaeologists, documented, and then stored away in a Mexico City museum. Hristov removed material from the head and sent it to the Max Plank Institute for Nuclear Physics in Heidelberg, Germany. After testing, scientists concluded that the material was 1800 years old. Hristov also had the figure evaluated by art experts. They concluded that it matches figures that were created by the Romans in about 200 A.D. Hristov and other scientists agree that the figure is solid proof that there were indeed Romans wandering about Mexico as early as the second century A.D.

Ships and sailing are documented early in the Old Testament. While these sailors didn't have all the tools we have today, they were just as intelligent as today's sailors. They were also filled with the same drive to explore and subdue the Earth as we are today, a drive that was placed in us when our forefather, Adam was created.

Prayer: Lord, help me to learn more about your world and praise you for it. Amen.

REF.: *The Province*, 2/10/00, p. A38, "Romans said first to find New World."

Ants Challenge Natural Selection

Genesis 6:19
"And of every living thing of all flesh two of every sort shalt thou bring into the ark, to keep them alive with thee; they shall be male and female."

Charles Darwin recognized that ants challenged his theory of natural selection. He even mentioned it in his book *Origin of Species*. He even asked how the situation with the lowly ant can ever be reconciled with his theory. He never did come up with an answer, and modern evolutionists still don't have any good answers.

Darwin's problem was with the worker ants. Even though they are products of sexual reproduction, they differ greatly from their parents. They are each specialized with features their parents don't have so they can carry out their designated tasks in the nest. The problem is that these workers are sterile females, so they cannot pass on the traits that are unique from their parents. Modern evolutionists theorize that perhaps there were some lucky mutations that took place in queen ants through their evolutionary history. However, this explanation is not very credible since the oldest fossilized ants are identical to today's ants. That means that there is no evidence of evolution in ants over a period of 70 million "evolutionary" years. Of course, we at **Creation Moments** don't accept claims that the world is that old.

Perhaps God, in His foreknowledge, designed ant society this way to foil Darwin and those who have thought like him throughout history. Whatever the case, here is evidence that the ant neither evolved nor could have possibly evolved.

Prayer: Dear Father, with wisdom you have designed your creation to bear witness to you. I, too, praise your Name. Amen.

REF.: *CRSnet*, 2/9/00, "Evolution, Sex and the Ant."

Patterns of Design

2 Timothy 1:13
"Hold fast the from of sound words which thou hast heard of me, in faith and love which is in Christ Jesus."

If you own a car, you know that it has an engine, doors, and, of course, wheels. There are some engineering solutions to transportation problems that are clearly better than others and no matter who makes the car, designers eventually arrive at the same solutions because they work best.

The same pattern holds true in creation. If you were designing genetic material that did one thing in one type of creature and a similar thing in a completely different type of creature, you would use the best design for both. That would result in similar genetic material in both creatures that was subtly different enough to do its unique job in each creature. As a result of using the best design solution, rather than chance, two very different creatures would have similar genetic material. And that's what we are finding as we learn more about genetics.

For example, we know that a particular gene in a developing mouse embryo influences the development of the back part of its brain. A very similar gene in the fly influences the development of its antenna, which are sensors that are tied into the brain. Another gene that is similar in people, fish and flies, influences the unique development of eyes in those creatures.

Similarities like these argue against the chance development of the genetic code. The same Designer that invented the genetic code also gave us His Word, which He also intelligently designed to bring us the knowledge of salvation in Jesus Christ.

Prayer: Lord, I thank you for bringing your saving Word into my life. Amen.

REF.: *Creation Research Society Quarterly*, 9/99, pp. 62-67, "Embryology and Evolution."

"Superb," "Splendid," and "Lovely"

Psalm 50:11
"I know all the fowls of the mountains, and the wild beasts of the field are mine."

There are thirteen species of a brightly plumed little songbird known as the fairy-wren. The birds are found in Australia and New Guinea. So colorful are their feathers that the various species go by names like "superb," "splendid," and "lovely." Even more noteworthy, however, is the birds' unusual behavior.

A male courting a female will bring her a flower petal. The petal usually matches his color or is a deeply contrasting color. Normally a perky little bird with an upright tail, when courting he lowers his tail and creeps around close to the ground. As he twists his body back and forth, he puffs out his cheek feathers. If the female accepts his courting, she builds their nest alone, lining it with bright parrot feathers.

While they mate for life, these birds are not known for fidelity to their mates. When mature, females will go off on their own, but males may stay with their parents for a year or more. Their main duty is to guard the family nest. If danger approaches the nest, the guard will puff up his wings, lower his tail and scuttle through dry grass, pretending to be a mouse. The idea is to lure the predator away from the nest.

The beauty and unusual behavior of these little birds testifies to more than God's creativity and love for beauty. They remind us of the beauty that was lost to God's creation when it was tainted by man's sin. Thankfully, some of that beauty that was lost can return to our lives through the forgiveness of sins that is found in Jesus Christ.

Prayer: Dear Father, I thank you for the beauty of your creation and for giving me the forgiveness of sins in Jesus Christ. Amen.

REF.: *Natural History*, 11/94, pp. 56-62, "Faithful Philanderers."

The Cruel God of Evolution

1 Corinthians 15:22
"For as in Adam all die, even so in Christ shall all be made alive."

Some people think that they can believe in Jesus Christ as their Savior and at the same time believe God created through evolution. These people are called theistic evolutionists or progressive creationists. What kind of nature would a god have who creates through the death that is essential to evolution?

First, death would have to have begun from the very moment of creation. Death, disease, pain and suffering would have had to originated with this god. Long before man, according to evolution, dinosaurs were eating other dinosaurs and entire species were dying out. Rather than loving us, such a god would be indifferent and capricious. The god that would use evolution to make living things would have no right to punish sin. He would have had no right to bring a world wide Flood to punish sinful man. This kind of god would be nothing more than a bully. Such a god would not be likely to have given us his word. And nothing the Bible says about him could be true. If the Bible really is from him, it is full of untrue claims. Any god who created using evolution is not the God of the Bible. If you think about it, the god of evolution sounds surprisingly like the devil.

What's wrong with believing that God created through evolution over millions of years? Any god that uses evolution and death to make us is extremely cruel. More importantly, if there was no first Adam to bring sin and death into the world, there was no need for Christ, whom the Bible calls the last Adam, to die and save us from sin and death.

Prayer: I rejoice, Dear Father, because you are gracious and merciful to me. Amen.

REF.: *Creation*, 9-11/99. pp. 42-45, "The god of an old Earth."

Your Multipurpose Eye

Scientists have learned that the eye is a multipurpose organ. They have long tried to figure out how our internal biological clock is set. This amazing clock is located in the brain and follows a period of almost exactly a day. It controls many of the body's rhythms, so it's important that it is properly set.

Scientists do know that the clock is set by light. They also know that your clock is not set by the rods and cones in your eye that see color and images. In fact they have conclusively shown that the eye's rods and cones have nothing to do with setting the clock. For a recent experiment researchers bred mice whose eyes lacked rods and cones. As they shifted light periods, the mice still adjusted their biological clocks to the new periods.

Now researchers are convinced they have found how our eyes adjust our biological clock after studying, of all things, the African horned frog. While studying the frog, scientists discovered a photoreceptor called melanopsin. This photoreceptor is found in the skin, eyes, and brain of the frogs. They then decided to test for melanopsin in our eyes. They discovered that a form of the photoreceptor is produced in the inner retina of our eyes. The rods and cones of our eyes are located on the outer retina. The human eye, and those of mammals, is a multi-purpose organ!

The intricate and precise ways in which we have been made increasingly glorify our Creator as we learn more about how we have been made. Surely it takes more faith to believe that such an elegant system as our biological clocks evolved by chance.

Prayer: I, too, glorify you, Lord, for I am fearfully and wonderfully made. Amen.

REF.: *Science News*, 2/19/00, p. 120, "Protein may help the eyes tell time."

Do You Know How Complex a Single Cell Is?

Psalm 96:3
"Declare his glory among the heathen, his wonders among all peoples."

It wasn't until 1839 that cell theory was first described by Theodore Schwann in basically its modern form. By 1858, researcher Rudolf Virchow had learned enough about the cell to conclude that every cell must come from a preexisting cell. But Charles Darwin wasn't paying attention. The very next year he published a book, *On the Origin of Species*, which theorized that the first cell was formed from non-living matter.

In Darwin's day, the cell was thought to be just a simple sac filled with jellied carbon. This concept is the origin of the term protoplasm. However, nearly 150 years of cell research has shown us that even protozoan and fungi cells are hugely complex. Today we know that even the simplest of these cells, eukaryotes, have an estimated 100,000 parts. Many thousands of different operations are taking place continuously within each of the cell's many parts, called organelles. Furthermore, the cell cannot live until all these parts are working properly. Even the simple E. coli bacterium has 4,000 genes. If we were able to magnify the DNA of the E. coli to the thickness of a clothesline, it would be five miles long!

There is no such thing as a "simple cell." The fact that a cell cannot live without all these thousands of parts shows that cells were created in their finished form, just as the Bible says. Even bacteria declare the glory of our Creator!

Prayer: Lord, I glorify you, for even the simplest cell testifies to your glory. Amen.

REF.: *CRSQ*, 3/99, p. 228, "The Putative Evolution of the Animal Eukaryote Cell Ultrastructure."

What Was the Monster of Troy?

I Corinthians 12:2
"Ye know that ye were Gentiles, carried away unto these dumb idols, even as ye were led."

People thousands of years ago may have "discovered" fossils without knowing what they had found. That's the new theory proposed by a group that includes a folklorist and some paleontologists. The theory is based on the figures painted on a 2,500-year-old Corinthian vase.

The colorful vase was made about 550 B.C. Called the Hesione vase, it shows the mythical woman Hesione being rescued from the monster of Troy by the Greek hero Herakles. Only the monster's head is shown emerging from a cave. The head looks very much like a giant skull with a tongue and a lizard-like eye socket. Those behind the new theory conclude that these details were added to what was really a very common fossil in the area.

Paleontologists say that the area around the Aegean Sea and western Turkey contains many fossils of large giraffes, camels and horses. Sometimes these fossils become exposed as the rocky hillsides weather. When the ancient Greeks found these large fossils, they thought that they were the bones of gods and monsters.

While this may seem superstitious to us today, we should be careful. Today's great superstition is that life evolved from non-living matter over millions of years. Belief in evolution is no less an attempt to escape acknowledging our Creator God than the legends of the ancient Greek gods. St. Paul himself wrote that the eternal power and divine nature of God are evident through the creation.

Prayer: Dear Father, I thank you for the light of your Word. Amen.

REF.: *Science News*, 2/26/00, p. 133, "Vase shows that the ancient dug fossils, too."

Biblical Skepticism Challenged

Psalm 119:160
"Thy word is true from the beginning: and every one of thy righteous judgments endureth for ever."

Unfortunately, many biblical archaeologists do not accept the Bible's account of the period of the judges. They believe that there was no distinct Israel during this time.

In a recent talk at Northwestern University, however, William Dever severely challenged these views. Dever is professor of Near Eastern archaeology and anthropology at the University of Arizona. He pointed out that during the period of the judges, about 300 new villages appeared out of nowhere in the central hill country of Palestine. The distinctly different design of the houses and villages match the living arrangements practiced in Israel at this time. Evidence, he noted, points to an increase in population that could only be accounted for by a rapid influx of people.

Furthermore, while pigs were a common staple at this time in Palestine, these new towns are unique in that they contain no pig remains. Farming also changed rapidly as hillsides were terraced. The introduction of iron affected daily life. And pottery styles changed rapidly as the culture developed. Moreover, an Egyptian artifact of the time conclusively proves biblical history. A monument erected in Egypt during the period of the judges mentions Israel as a distinct people.

While science can illustrate many details found in the Bible, we don't believe what the Bible says because of science. Therefore, science can never "prove" the Bible to be in error. Faith convinces us of the Bible's truth.

Prayer: Dear Father, I rejoice for the faith you have given me that your Word is without error. In Jesus' Name. Amen.

REF.: *Biblical Archaeology Review*, 3/4/00, pp. 28-35, 68, "Save Us From Postmodern Malarkey."

Calibrating the Bee Odometer

Psalm 31:3
"For thou art my rock and my fortress; therefore for thy name's sake lead me and guide me."

When a honeybee finds a good source of food she returns to the hive to tell fellow bees where it is. Her dance tells nest mates the direction of the source from the nest, and how far away it is. Researchers have long puzzled over how the bee knows the distance she has traveled. Several theories have been proposed, but until now, none has been proven.

Recently, an international research team has convincingly demonstrated how bees know how far they have traveled visually. Researchers trained bees to fly through a 30-foot tube by placing food at the far side. Then they placed a checkerboard pattern inside the tube designed to make the bees think they had flown further than they really had. Researchers found that the pattern made the bees think they had flown 31 times further than they really had. After watching the bees that had gone through the tricky tube, researchers calculated that for every 18 degrees an image travels across the bee's eye, she will dance for one millisecond as she communicates her discovery. As one researcher put it, they have calibrated the bees' odometer.

Bees themselves are a wonder of design. The way they communicate their finds with one another are even more difficult to explain by those who accept evolution. These are all designs of the Creator who wishes to guide us to Himself. He communicates with us through His Word that tells us of salvation through the forgiveness of sins that is found in His Son, Jesus Christ.

Prayer: Lord, continue to teach me to understand and believe your Word. Amen.

REF.: *Science News*, 2/5/00, p. 87, "Bees log flight distances, train with maps."

Did Ants Invent the Electric Knife?

Romans 8:28
"And we know that all things work together for good to them that love God, to them who are the called according to his purpose."

Scientists have discovered that seemingly purposeless behaviors among leaf-cutting ants actually serve an important function. Leaf-cutting ants cut green leaves into pieces that they can carry back to the nest. As they cut a leaf, they vibrate, creating an effect like an electric knife. Since they can cut the leaves without vibrating, it would seem to have no purpose. Smaller worker ants accompany the larger worker ants to their leaf-cutting site. Then they wait around, watching the larger workers. To make matters stranger, once a leaf section is ready to be carried back to the nest, the little do-nothing workers climb aboard the leaf to be carried back home with it.

New research has revealed that these apparently pointless behaviors serve a single, important purpose. As the larger worker finishes cutting a leaf section she vibrates even faster. The increased rate of vibration is apparently a signal to the smaller worker to board the leaf section for the ride home. But why are the smaller workers there in the first place? Researchers have discovered that they discourage attacks by phorid flies. These flies would otherwise inject an egg into a leaf-cutting ant. When the egg hatches, the larva chews its way through the ant, killing it.

This amazingly designed system reminds us that there is purpose throughout the creation. God is the Author of purpose, not mindless chance. God's highest purpose for us is to come to faith in Jesus Christ as our Lord and Savior.

Prayer: Lord, I thank you because you have brought purpose into my life. Amen.

REF.: *Science News*, 2/5/00, pp. 92-94, "When ants squeak."

Roman Artifacts Discovered in Brazil

Acts 20:13a
"Then we went before to ship and sailed unto Assos, there intending to take in Paul..."

Both Spain and Portugal claim to have discovered Brazil in the 16th century. However, mounting evidence suggests that neither country can claim discovery of Brazil.

Guanabar Bay is on the Brazilian coast, less than ten miles from Rio de Janeiro. In 1976 a diver discovered two unbroken amphoras at the bottom of the bay. Amphoras were tall storage pots that were commonly made and used by the Romans. In 1982 an archaeologist discovered thousands of fragments from still more Roman amphoras in the same area. Among the fragments were 200 amphora necks. The styles of these amphoras indicate that they were made in second century A.D. Rome.

Ancient Roman shipwrecks have also been found in the Azores islands off the west coast of Spain. The Azores would have been a good European starting point for crossing the Atlantic. The shortest route across the Atlantic from the Azores lands one on the coast of Brazil. Modern sailing ships make the crossing in only 18 days. So even before the amphoras were found, it would not have been unreasonable to suppose that the Romans, who were skilled sailors, had made the crossing. The discovery of the amphoras proves that at least one Roman ship made the trip.

The belief that up to Columbus' time people assumed the world was flat is a myth. God created intelligent and curious human beings with the ability to explore and learn about God's creation.

Prayer: Thank you, Lord, for allowing me to appreciate your creation. Amen.

REF.: *Science Frontiers*, p. 25, "Romans in Rio?"

Get a Younger Brain!

Genesis 5:32
"And Noah was five hundred years old, and Noah begot Shem, Ham, and Japheth."

How did Noah have sons and build the ark when he was 500 years old? Today, scientists can measure a marked decline in mental abilities even in a healthy 50-year-old man. Undoubtedly there are several factors, with God behind each of them. One of the answers appears to be nutrition.

Research done by Tufts University showed that middle-aged men with the lowest amount of vitamin B6 in their blood only scored half as well as men of the same age with the highest level of vitamin B6. Another four-year study looked at a population of people in which 15 percent could be expected to develop Alzheimer's disease. It found that not one person who took vitamin E or C developed the disease.

Similarly, a Framingham Heart study found that people who ate three extra servings of fruits and vegetables a day reduced their stroke rate by 22 percent. And when University of New Mexico researchers looked at the effects of vitamin B supplements in older people, they too discovered a nutritional link. Those who took the supplements not only scored higher on mental functions than those who did not, they did as well, or better than young people. Likewise, over 50 studies have confirmed that ginkgo biloba helps memory and concentration, reduces confusion and mitigates effects of Alzheimer's disease.

Our wise Creator has given us powerful nutritional tools to help our earthly time be healthy and productive. Perhaps Noah and the other long-lived ancients knew more about how to eat healthily than we do.

Prayer: Lord, make my life here healthy until you call me home to yourself. Amen.

REF.: *USA Weekend*, 3/5/00, p. 7, "Important News your brain can use."

What's Part Mammal, Marsupial and Reptile?

Psalm 111:4
"He hath made His wonderful works to be remembered; the LORD is gracious and full of compassion."

While all living things defy evolution, some do it more clearly than others. The echidna is one example of a creature that obviously challenges evolution. This Australian marvel is often called the spiny anteater. It has little resemblance, however, to anteaters in other parts of the world.

Echidnas are classified as monotremes, which are egg-laying mammals. While the adults have no teeth, a hatchling echidna escapes from its egg with an egg tooth. Adults have a long snout and an even longer sticky tongue that catches ants. The hatchling is protected and nursed in a marsupial-like pouch. As it grows it develops the long spines of an adult. If the echidna sounds like a strange creature, its mating habits are even stranger. At mating time echidnas form "trains." A large female leads the train, followed by up to seven males, the smallest being the last car of the train. They walk single file until the female finds a tree she likes. Then the males dig a trench around the tree and they seek to drive each other out of the trench. Only the victorious male is allowed to mate with the female.

With its mammalian, reptilian and marsupial features, the echidna challenges evolution. Evolutionists would have a hard time explaining what forces would produce a creature that is apparently related to no other. The best explanation for the echidna is that it is the work of an infinitely creative and powerful Creator.

Prayer: Dear Father, Your unlimited creativity fills me with joy. Amen.

Ref: *Science Frontiers*, p. 129, "Echidna Eccentricities."

191

New Antibiotic Discovered Under the African Jungle

Matthew 8:3
"And Jesus put forth his hand, and touched him, saying, 'I will; be thou clean.' And immediately his leprosy was cleansed."

Researchers searching for new medicines having been scouring the jungles of the world. Many of our new and most powerful drugs have come from Old and New World jungles. Sometimes researchers learn something about these medicines from local healers, who already seem to know how to use them.

One of the most recent discoveries took place in the jungles of the African continent. Researchers from the University of Lausanne in Switzerland found out that traditional medicine practitioners used a yellow substance to treat syphilis and leprosy and to kill termites. They learned that the substance comes from the root of a tree that is common across Africa's jungles. The researchers collected some samples to see if the yellow substance might actually have compounds medicine could use. They found a compound that kills microbes and fungi. The scientists concluded that the compound is manufactured by the tree to protect its roots from dangerous fungi in the soil. Now, they are going to have to learn to make the compound in the laboratory. Researchers say that six trees would have to be killed to harvest just 50 grams of the new antibiotic.

Our merciful God knew that we would fall into sin and become subject to illness. So when He made the creation, He gave us many plants that can provide medical help if we explore that creation as He commanded. Still these medicines don't treat the underlying cause of our problems - sin. So He sent His Son to save us from the guilt of our sin.

Prayer: Lord, I thank you for earning forgiveness and salvation for me. Amen.

Ref: *Science News*, 3/4/00, p.159, "Rooting for new antimicrobial drugs."

Eating that Can Result in Starvation

Psalm 102:5-6

"By reason of the voice of my groaning my bones cleave to my skin. I am like a pelican of the wilderness; I am like an owl of the desert."

The creosote shrub is common in the desert of the American Southwest and northern Mexico. This remarkable plant is not only well designed for life in the desert, but also it protects itself from those who would munch its leaves.

Creosote bushes grow in stands because they reproduce by sending shoots out from a collective root system. One stand measured over 60 feet across at its widest point! Stands can be hundreds of years old. The bushes produce several hundred chemicals that evaporate into the air. Stands are said to smell like chemistry laboratories. Other desert plants avoid growing near them, probably because of the smell. As a result, stands do not have to share scarce water supplies with other plants. Compounds call phenols and phenoloxidasa are among the chemical produced by the leaves. These chemicals must be stored in different compartments within the leaves so that they do not react with each other. However, when the leaves are chewed, these compartments break down. They react in the stomach, producing chemicals called quinones. The quinones then react with the plant proteins, changing them into a form that cannot be digested. The result is that whatever tries to eat creosote leaves can have a full stomach and be starving at the same time!

The creosote's advanced chemical warfare could not possibly be a product of mindless evolution. Its methods of protection reveal a well thought out design that can only be credited to a wise Creator.

Prayer: Dear Father, in Christ I flee to you from the desert of my guilt. Amen.

Ref: *Bombardier Beetles and Fever Trees*, pp. 25-27.

Solar System Structure Confounds Scientists

Our solar system has a nice, neat structure. The planets all have nice stable, relatively round orbits. This structure protects the Earth from being struck by another large planet. Evolutionary theories about the origin of the solar system have always said that it formed into its current neat structure from a swirling disk of hot gases that once orbited the sun.

Now, with the discovery of planets orbiting other stars, astronomers are rethinking their theories about the how the solar system formed. The problem is that none of the other planetary systems have an orderly structure and all the planets so far discovered are closer to their stars than we are to the sun. The coolest among them has a surface temperature of 180° F. Most of these planets have orbits so erratic that collisions with other planets are always a threat.

So, astronomers wonder, how did our solar system become so well ordered for life? They argue that the large icy planets, Uranus and Neptune, are too large to have formed at the edge of the solar system. So it is concluded that they must have formed closer to the sun, meaning that the solar system was once as chaotic as the recently discovered planetary systems.

This new evolutionary approach simply introduces another problem: How could our present, orderly solar system have developed out of the chaos of a system in which planets are crashing around like billiard balls? The Bible still offers the best answer. The solar system was designed and created for us by a wise and loving God.

Prayer: I rejoice, Dear Father, because the power you bring to me is your love. Amen.

Ref: *Discover*, 3/00, p. 54, "Solar Revisionism."

Archaeology Sheds New Light on Israel's High Places

II Chronicles 11:15

"And he ordained him priests for the high places, and for the devils and for the calves which he had made."

The Old Testament frequently, and disapprovingly, mentions the fact that the Israelites often built "high places." The Bible indicates that high places were worship centers where a pagan religion or a mix of Israelite and Canaanite religion was practiced.

Until recently, high places found by archaeologists revealed little about these sites of false worship. The discovery of a high place at Rehov, however, tells a much more complete story—one consistent with what the Bible tells us. Its dating could make this high place one of those for which Rehoboam, son of Solomon, appointed priests. The sanctuary was a raised ten-foot square of mud brick. In the middle was a stone platform that was three feet square. One side of the platform had four large stones, two of them taller than the others. These may have been the "standing stones" mentioned in the Bible. In front of the platform was an offering table. Many of the bones of offered animals, often wild goats, were found at the site. Such offerings would have been against God's law. In addition, various figurines were found at the site. These included a bull, two female figures and another animal. Israel was forbidden to make such images.

This discovery challenges the skeptics and upholds the integrity of the Biblical history. Once again, the Bible's history has proven to be factual, even when reporting Israel's sins.

Prayer: Lord, preserve me from being misled in any way from your truth. Amen.

Ref: *Biblical Archaeology Review*, 3-4/00, pp. 38-51, 75, "Will Tel Rehov Save the United Monarchy?"

The Deadliest Animal in the World

I Peter 5:8
"Be sober, be vigilant; because your adversary the devil as a roaring lion, walketh about, seeking whom he may devour."

You have probably never heard of the last animal you would ever want to meet undefended. This animal, found in the wild only in Madagascar, is called the fossa (pronounced FOO-sa). A typical specimen weighs only about 14 pounds. Its body is about 28 inches long, but its tail is another 28 inches long.

The fossa looks something like a small mountain lion and was originally classified as a member of the cat family. But it also has features and behaviors, including its tenacity in fighting, more similar to the mongoose. This led scientists to reclassify the fossa as a member of the mongoose family. These ferocious creatures live in the forests of Madagascar. Despite its small size, it is said to eat anything with a heartbeat. This powerful creature is known by science to bring down even wild pigs much larger than itself. The local people of Madagascar say it can even bring down oxen. Pound for pound, it is recognized as the deadliest animal in the world. Scientists have also learned that an even larger version of the fossa possibly existed 1500 years ago when Madagascar was first settled. This fossa weighed as much as 200 pounds and had a body 6 feet long, not counting the tail. It could have brought down any creature it wanted.

While the fossa may be considered the personification of evil, the Bible tells us that the devil is behind all evil and is much more dangerous than any fossa. So we can thank God that His Son, Jesus Christ, has won for us the victory over sin, death and the devil.

Prayer: Lord, I thank you. You have given me your victory over the devil. Amen.

Ref: *Discover*, 4/00, pp. 68-75, "The deadliest carnivore."

Secular Studies Bolster Marriage Commitment

Proverbs 18:22
"Whoso findeth a wife findeth a good thing, and obtaineth favour from the LORD."

Hundreds of studies have shown that married couples enjoy a happier, healthier life than cohabiting couples. And each study supports the same conclusions: cohabitation without the commitment of marriage is a formula for trouble in life.

A UCLA researcher evaluated over 130 studies on the happiness and health of married couples versus cohabitors. He concluded that married couples scored much higher in personal happiness than cohabitating couples. They also had much lower rates of suicide, alcoholism, psychiatric care and general health problems. But there is more: A Washington University study found that cohabiting couples have less healthy relationships than married couples. Then, a study released in 1999 by the National Marriage Project reviewed ten years of studies and concluded that cohabiting couples who eventually marry greatly *increase* their chance for divorce. A 1997 study by the National Survey of Families and Households found that couples who lived together before marriage were 46 percent more likely to get divorced than those who do not.

The Bible recognizes that not everyone is cut out for marriage. No one, though, is cut out for cohabitation. We need the commitment of marriage. Modern science is discovering that the Bible was right all along in saying that a man and a woman need the commitment of marriage if their union is to be blessed.

Prayer: I thank you, Dear Father, for the gift and blessings of marriage. Amen.

Ref: *Christian News*, 3/6/00, p. 10, "Science, Scholarship and Scripture."

197

The Increasingly Tangled Tree of Life

John 1:3
"All things were made by Him; and without him was not any thing made that was made."

Researchers studying genetics have concluded that the evolutionary tree of life does not reflect the facts of biochemistry.

By the 1960s, biochemists had concluded that living things could be grouped into two distinct types based on their basic structures and genetic information. Eukaryotes have one or more cells and a true nucleus. Prokaryotes have smaller cells with no true nucleus. They are so different from eukaryotes that researchers concluded that they must have developed separately from non-living matter. By the late 1970s a third type of life was recognized, the archaea. Archaea favor extreme environments such as undersea vents. Their biochemistry is unlike prokaryotes or eukaryotes.

According to evolutionists life may have had to develop from non-living material three times. Further, to explain the biochemistry of these three types of creatures biochemists have to assume that at points these separate types had to exchange genetic material with each other, and with a fourth, unknown and extinct type of living thing!

One logical conclusion of this new evolutionary approach is that life had to arise from non-life many times. Yet, modern biochemistry still has not explained how life could arise from non-life even once. And, this is exactly the pattern we might expect if all living things are the product of a wise Creator Who used similar designs to solve similar problems.

Prayer: Dear Father, great and glorious is the work of your hands! Amen.

Ref: *Scientific American*, 2/00, pp. 90-95, "Uprooting the Tree of Life."

High-Tech Ants

Genesis 1:31
"And God saw everything that he had made, and, behold, it was very good. And the evening and the morning were the sixth day."

Why do ants like electricity and why can they sense microwaves? If evolution were true, how would ants evolve a love for electricity in only the short time that man has been producing it? For what purpose would they have evolved the ability to sense microwave radiation long before man was using it?

It has been well established that ants like to build their nests around electrical equipment. They are often found around electrical meters. They are famous for building their nests around airport runway lights. They seem to be particularly fond of electrical relays. No one knows why. Now we learn that unlike any other multicelled creatures, they can survive inside a running microwave oven, perhaps enjoying your food as it cooks! How can they do this?

It seems that ants can detect the invisible wave pattern of microwaves inside the oven. These patterns leave some areas of the oven low in microwave radiation while other areas are high in radiation. Ants simply avoid those areas where the radiation is strong enough to kill them. Incidentally, you can see these patterns by lining the tray on the floor of your microwave oven with marshmallows. You will notice that some will become very cooked while others are barely touched by the microwaves.

If evolution were true, how would ants develop the ability to deal successfully with conditions that did not exist? It makes more sense to conclude that our Creator, Who knows the future, made the ants as He did for reasons we still must discover.

Prayer: Lord, I am filled with wonder at your wonderful work. I thank you for all you have given me. Amen.

Ref: *Science Frontiers*, 1-2/00, p. 2, "Ants Like Microwaves."

Ancient Meteorologists Predicted El Niño

Genesis 1:14
"And God said, 'Let there be lights in the firmament of the heaven to divide the day from the night; and let them be for signs and for seasons, and for days and years..."

El Niños have been periodically changing the weather patterns of North and South America for thousands of years. El Niños in the Andes result in a four to six week delay in the rain needed for the potato crops on which ancient farmers depended. They even knew ahead of time when to wait to plant their potatoes for best results.

Between June 13 and 24 of each year these farmers would search the early dawn sky for the invisible signs of an El Niño. The warm El Niño waters cause more moisture to accumulate in the atmosphere. Some of this moisture forms invisible cirrus clouds high in the atmosphere. While invisible, these clouds would cause some of the more dim stars to disappear. So the Andean farmers looked to the northeastern horizon to find the Pleiades. If all seven stars, including the dimmest of them, shone clearly morning after morning, the Andean farmers knew that no El Niño was forming. But, if they consistently saw a dimming of the stars, they knew that they should delay planting for four to six weeks because the rains would be late.

Where would the Andean farmers learn the signs and consequences of an El Niño for their crops? The Bible teaches that the sun, moon and stars also serve to give us signs about the seasons. It would appear that God originally gave Adam the ability to read these signs, knowledge which Adam would have passed down to his future generations.

Prayer: I praise you, Lord, for the entire creation shows your handiwork. Amen.

Ref: *Science Frontiers*, 3-4/00, p. 1 "Archeometeorology."

The Mystery of Mouse Pyramids

Psalm 118:22
"The stone which the builders refused is become the head stone of the corner."

Mice of the Sahara Desert pile up round stones in front of their burrows making small pyramids through which the air can flow. As the morning air heats up, the rocks heat up much more slowly. As the dew point of the rapidly warming air rises, water in the air begins to condense on the cooler stones. It turns out that having round stones leaves exactly the right shaped spaces for this to take place. This is how the mice get their water.

Mice in the dry climates of Australia also build pyramids to catch the morning dew. They construct their pyramids of small, uniformly sized round stones over their burrows. Their pyramids may be up to a yard across. Again, as the morning air passes through the loose pile of stones, the dew collects on the cooler stones, providing a source of water for the mice. While a second species of Australian mice doesn't build such pyramids, it does use those built by the first species as a source of water.

All of this leaves those who believe in evolution with some difficult questions. How did mice learn to build these pyramids to catch the dew? And how did two different species, half a world apart, arrive at the same solution to survival?

These questions are easily answered if we assume that the intelligent Creator of mice built such knowledge into them when He first made them. This is the same Creator Who sent His Son, Jesus Christ, to be the cornerstone of the eternal structure made up of all believers.

Prayer: Dear Lord Jesus Christ, thank you for being the cornerstone of my life. Amen.

Ref: *Science Frontiers*, p. 17, "Ancient Greek Pyramids?," p. 137, "More Mouse Engineering."

The Genius of Insect Flight

Psalm 105:31
"He spake, and there came divers sort of flies, and lice in all their coasts."

Flying insects have been a huge mystery to scientists, especially those who believe in evolution. Until recently, scientists didn't know how flying insects could fly. Wind tunnel tests on insect wings showed that their wings produce anywhere from a third to half as much lift needed for flight, depending on the species. Yet horseflies have been clocked flying at up to 90 miles per hour!

Detailed research on the biomechanics of insect flight has now revealed how they manage to fly. Insect flight does not use quite the same principles as airplane flight. Unlike the wings of a modern airplane, insect wings trace a figure eight during flight. If an airliner's wings attack the air at an angle of more than 18 degrees, the vortex that provides lift to the top of the wing pulls away from the wing and the result is a stall. Insect wings, however, attack the air at a much steeper angle. Why doesn't the lifting vortex pull away from an insect's wing, leaving the insect in a stall? Researchers learned that because the insect's wing is moving faster at the tip, the lifting vortex does not pull away from the wing and gives the insect a 70 percent increase in lift.

Evolutionists depict today's flying insects as descendants of ancient insects that could only glide. Then, over supposed millions of years they developed the wings and flight skills of modern insects. This recent research, however, reveals that insect flight is no chance development. It was carefully engineered by a Creator from the very beginning.

Prayer: I thank you Lord, for your mighty Word has created faith in me and brought me salvation. In Jesus' name. Amen.

Ref: *Discover*, 4/00, pp. 27-28, "What's the Buzz?"

The African Priestly Tribe of Israel

John 10:16
"And other sheep I have, which are not of this fold: them also I must bring, and they shall hear my voice; and there shall be one fold and one shepherd."

There are several groups of people around the world that claim to be part of the Jewish people. Many even practice a form of the Judaic rites. But they don't have any provable connections to the Jewish people.

In southern Africa there is a tribe called the Lemba. Lemba tradition says that the tribe was led to Africa from Judea by a leader named Buba in the distant past. The Lemba also practice Jewish traditions, including keeping one day a week holy. Also, they do not eat pork. This fact could be dismissed as coincidental, but the Lemba have a proven heritage. In a previous devotional we investigated the so-called "cohen genetic signature." (Cohen means priest.) This genetic mutation is carried in the Y chromosome and is rare in the general population. Even among the general Jewish population, only three to five percent carry the mutation. The mutation is so common among the descendants of Aaron, however, that it is considered a strong indicator of descendancy from the Levites. It is also common among the Lemba men, proving they are not only genetically Jewish, but also that they descended from ancient Israel's priests!

While we may never know exactly how the Lemba tribe came to be, we know that all people are God's creation. We also know that Jesus Christ is the Savior of the nations. No one has wandered so far from God that he or she cannot be saved by the blood of Jesus Christ.

Prayer: Dear Father, with a rejoicing heart I praise you for your love to me. Amen.

Ref: *Science Frontiers*, 1-2/00, p. 1, "A Far-Wandering Tribe?"

A Prehistoric Spark Plug?

I Corinthians 10:4b
"For they drank of that spiritual Rock that followed them, and that Rock was Christ."

"Creation Moments" frequently reports on fossil finds that show that rocks don't take millions of years to form. The program features many of these, rather than just one example, in the hope that it will become common knowledge that it is not unusual for rocks to form rapidly.

Our latest example was found on a beach in Ventura County, California. It is clearly a modern rock that formed in the last few decades, because partially embedded in the stone is a spark plug. Clearly, no one would claim that man has been making spark plugs for millions of years.

The porcelain part of the spark plug became part of the rock. Originally, the metal parts of the spark plug were also embedded in the rock. They rusted away after the rock was formed, leaving an imprint clear enough to determine the thread size of the now missing metal parts. This suggests that this rock formed within a relatively short time, since such metal quickly corrodes in seawater. Based on the thread size and the porcelain portion of the plug, it is thought that the spark plug doesn't date to before the 1950s! It is theorized that the plug originally was in a boat engine and was discarded at sea.

That there are so many examples of obviously modern fossils shows that it is not uncommon for rocks to form rapidly. Some rocks you see today may not have existed as rocks when your grandfather was your age. There is no reason for the long-age dating of rocks to undermine faith in our spiritual Rock, Who is Christ.

Prayer: Dear Father, thank you for sending your Son to be my Rock. Amen.

Ref: *Creation*, 9-11/99, p. 6, "Sparking interest in rapid rocks."

Take Your Chocolate Medicine

Romans 5:9
"Much more then, being now justified by his blood, we shall be saved from wrath through him."

Chocolate has long been thought of as a junk food. New research on several fronts now reveals that chocolate should probably be reclassified as a health food.

Research shows that the beans from which chocolate is made contain large amounts of antioxidants called flavonoids. Found in tea and other foods, these substances have been linked to a reduced risk of cardiovascular disease. These flavonoids are present in high enough quantity in the chocolate you buy at the store to produce measurable results in scientific studies. A single gram of chocolate milk has 10 milligrams of antioxidants. If you like dark chocolate, you will more than double your dosage of antioxidants.

While some 4,000 natural flavonoids have been identified, the types found in chocolate are among the most powerful known. They are even more effective than the antioxidants found in vitamin C! Other flavonoids in chocolate have been found to relax the inside of blood vessels that lowers high blood pressure. Chocolate flavonoids have also been found to work as a mild aspirin would to help thin blood. They also help keep the platelets from breaking, which happens when platelets get too sticky. As if all these benefits aren't enough, chocolate has also been found to raise good cholesterol.

Chocolate may help you have a healthy cardiovascular system in this life. But when this life is done, only the blood of Jesus Christ will give you eternal life.

Prayer: Lord, I thank you that you shed your blood for me. Amen.

Ref: Science News, 3/18/00, pp. 188-189, "Chocolate Hearts: Yummy and Good Medicine?"

The Bible's Amazingly Accurate Account of the Flood

Genesis 8:7
" And he sent forth a raven, which went forth to and fro, until the waters were dried up from off the earth. "

On May 18, 1980, Mount St. Helens exploded, devastating over 200 square miles in southwestern Washington State. The heat spewing from the volcano melted the snow that covered its heights, creating mudflows that covered the surrounding area. The grim prediction was that forests and wildlife would never be found on or near the mountain within our lifetimes.

Mount St. Helens' recovery from the devastation provides us with a small-scale model of the recovery from the great Flood at the time of Noah. Only ten years after the eruption, much of the plant and animal life in the devastated area had returned. Today, evergreens stand taller than a man, fish swim in the lakes and rivers, and frogs are in the ponds. Even larger animals like elk can be found in the once desolate area. This suggests that it would not have taken generations for the foliage that God preserved in the Flood waters to regrow forests and jungles, providing homes for the growing, migrating population from the Ark. Before that could happen, Noah had to know that there was somewhere for the animals to go. So he first sent out a raven. Mount St. Helens illustrates why Noah decided to send a raven first. Among the first new colonizers at Mount St. Helens was the raven. They are part of God's army of scavengers to remove the dead and can live in a wide variety of conditions.

Here again we see that even the small details recorded in the Bible are true to real life.

Prayer: Dear Father, I thank you for giving us your perfect Word of Life. Amen.

Ref: *Creation*, 3-5/00 pp. 33-37, "After Devastation, The Recovery."

Don't Make a Monkey Out of Me!

Romans 1:22-23

"Professing themselves be wise, they became fools, And changed the glory of the uncorruptible God into an image made like to corruptible man; and to birds and four-footed beasts and creeping things."

Scientists who believe in evolution have a faith just as much as any Bible-believing Christian. However, it is a very different faith than Christianity. This fact was illustrated when some secular scientists declared the bones labeled "Piltdown man" an ape-like human ancestor. Only after a couple of generations did they bother to actually study the bones and discover them to be an obvious hoax. They did the same with "Nebraska man," which turned out to be a pig.

Recently, scientists announced the finding of some fossil bones of the very first primate. Naming it E-o-sim-ias or "dawn monkey," they say that it led to monkeys, apes and finally to us. Eosimias weighed less than an ounce and could have stood on your thumb. Scientists found no complete skeletons. However, after examining foot bones the size of grains of rice, and some lower limb bones they declared it to be our ancestor. They interpreted the foot bones to be primate-like. They concluded that the lower limb bones show that Eosimias had grasping hands and feet that enabled it to walk on branches like monkeys.

The Bible warns that God will curse those with ever more foolish notions who reject the obvious fact that He is Creator. He does this so that they might see their foolishness and repent for placing the creation over the Creator. In this case they have done that by saying that it was not the Creator, but rather this tiny monkey that led to humanity.

Prayer: Lord, please continue to enlighten me with your wisdom. Amen.

Ref: *International Herald Tribune*, 3/17/00, "Tiny Fossil Fills Gap in Evolutionary Record."

Seahorse Monogamy, Sort Of

Matthew 19:6
"Wherefore are no longer twain, but one flesh. What therefore God has joined together, let no man put asunder."

A few years ago, a "Creation Moments" program highlighted the amazing fact that male seahorses, not the females, hatch the young and are solely responsible for their nurture. New research now tells us more about the relationship between the male and female seahorse.

One of the surprises from the research is that the female seahorse practices a limited form of monogamy. An individual male seahorse will range over only about a square yard of area. His mate, however, will range over 100 times that area. Although she isn't much of a homebody, she remains faithful to her mate during his entire pregnancy, greeting him each morning by joining him and flashing different colors.

It turns out that this greeting ritual is very important. In the laboratory, researchers placed one female and two males in the same tank. After the mating ritual the female deposited typically 200 eggs in the egg pouch of the male she selected. He was then moved to another tank until his two week gestation was completed. During the entire time, the female greeted the remaining seahorse each morning, but no attempts at mating took place. But, when the male she selected for mating was returned to the tank and the mating ritual began again, she always chose the male she had greeted each morning during the other's absence in the birthing tank.

Monogamy is not a product of evolution or human culture. It is a gift of God, Who created seahorses, people and all things.

Prayer: Thank you, Lord, for the gift of marriage. Help me to honor it. Amen.

Ref: *Science News*, 3/11/00. pp. 168-170, "Pregnant and still macho."

No Heartbeat? Don't Worry

I Chronicles 16:24
"Declare his glory among the heathens, his marvelous works among all nations."

Evolutionists say that an animal's unique features developed because the unique feature gives the animal some survival benefit. Just what that benefit might be is open to interpretation, and sometimes the explanation is a little fuzzy. Sometimes the explanation just doesn't work.

Such is the case with the marine iguanas of the Galapagos Islands. These iguanas are excellent swimmers and search underwater for much of their food. When they are underwater, though, they are vulnerable to the marine food chain. Sharks find the marine iguanas very tasty. With their sensitive hearing, these predators can hear the heartbeat of a hiding iguana 12 feet away. So part of the iguana's defense is to stop its heart beat voluntarily. Amazingly, the iguana can keep its heart stopped for up to 45 minutes without suffering any ill effects.

Evolutionists explain that this ability evolved because it gives the iguana a survival advantage in shark-infested waters. Obviously, a creature's ability to stop its heart for this long requires some major internal modifications. To believe that the iguana knew all the changes it needed to intentionally stop its heart for 45 minutes without death requires quite a leap of faith! Moreover, this evolutionary explanation just doesn't work. The Komodo dragon can also stop its heart. For the Komodo this ability provides no survival advantage, for these huge lizards have no natural enemies.

These lizards received their special abilities from their wise Creator. Their special abilities glorify Him, not mindless evolution.

Prayer: Dear Lord, Your works fill me with wonder. I glorify you as my Savior. Amen.

Ref: *Science Frontiers*, 3-4/00, p. "Heart-Stoppers."

Comparison Places Humans Closer to Chickens than Mice

Psalm 59:8
"But thou, O LORD, shalt laugh at them; thou shalt have all the heathen in derision."

According to evolutionary theory, humans had a common ancestor with chickens 300 million years ago. Much more recently in evolutionary time, we had a common ancestor with the mouse. One would then be led to conclude that, in evolutionary terms, we are genetically closer to the mouse than to the chicken. Of course, as Christians we have good reasons to reject all those multiple millions of years.

As they learn more about the genetics of various kinds of living creatures, evolutionists are receiving some surprises. Researchers often find that the long-supposed evolutionary relationships are contradicted by the genetic information. For example recent research compared the gene segments of chickens, mice and people. If evolution were true, one would expect more similarities between the genetic segments of mice and humans. To their amazement, researchers found that, of mice and chickens, we share more common genetic segments with chickens! They tried to explain this result by saying that perhaps mouse segments change more rapidly than human or chicken segments.

Again, new genetic knowledge has challenged evolutionary assumptions. Once we know what those genetic segments that were studied do, we might understand why these similarities exist. We predict that such knowledge will glorify our Creator's wise design.

Prayer: Lord, help me see more clearly how new knowledge glorifies you. Amen.

Ref: *Nature*, 11/25/99, pp. 411-412, "The dynamics of chromosome evolution in birds and mammals."

Jupiter Defies Naturalistic Explanation

I Corinthians 15:41
"There is one glory of the sun, and another glory of the moon, and another glory of the stars; for one star differeth from another star in glory."

According to evolutionary naturalism, the solar system began as a disk of hot, rotating gas. The sun began to form first. At the same time, the planets began to form as they collected material out of the gaseous disk. This process generated more heat, with the largest planets collecting more of the heat than smaller planets. The amount of heat collected by the large, heavy planets like Jupiter compared with a planet like the Earth accounts for the different elemental composition of the planets.

When scientists began studying Jupiter, they expected it to have certain lighter elements in the same proportion as the sun. Large planets like Jupiter would have been too hot at formation for lighter elements to condense. But, researchers using the Galileo space probe have discovered that Jupiter has more than twice as many gases as the sun. It has nearly three times as much krypton and xenon gas as the sun. And, Jupiter has over three-and-a-half times as much ammonia as was expected.

Scientists are trying to figure out where these gases came from. Evolutionary scientists admit that they cannot find a satisfactory naturalistic explanation for the chemical composition of Jupiter. They have even speculated that they might find the same problems with the other outer planets.

That the planets differ in composition is exactly as stated in the Bible " or one star differeth from another star in glory". I Cor. 15:41. That these differences don't fit evolutionary theory further witnesses to God's glory as Creator.

Prayer: Dear Father, help me glorify you along with the heavens. Amen.

Ref: *Nature*, 11/18/99, pp. 269-270, "A low-temperature origin for the planetessimals that formed Jupiter."

The Super-Supercomputer that Heals Itself

Isaiah 57:19
"'I create the fruit of the lips: Peace, peace to him that is far off, and to him that is near, saith the LORD; and I will heal him.'"

A healthy person with a minor injury heals without any thought required for the healing process. More severe injuries may require medical attention, but no amount of thought concentration on the healing process will speed healing along. This automatic healing process could not have evolved because without the power to heal in the first place, any injury to any creature ultimately would be fatal.

IBM is now working on a super-supercomputer that would run 500 times faster than any computer known today. It will be used to produce three-dimensional models showing how proteins fold as they do their work. It is hoped that the models will help medical researchers design more effective medicines. The computer system working on this task will have over one million processors linked to adjacent processors. The adjacent processors will monitor the main processors. If one of the main processors becomes faulty, the adjacent processor will remove it from the computing system electronically. If it works, the entire system will be known as a self-healing computer.

While the computer may be able to heal itself, it will have to "think" about healing constantly. And despite its speed, its healing and computing abilities are still slower than ours.

God has given us great powers of healing. But we can do nothing to heal ourselves spiritually. For that we must rely on Jesus Christ and the salvation He brings.

> *Prayer: I thank You, Lord, for the healing of my soul through forgiveness. Amen.*

REF: *New Scientist,* 12/11/99, p. 8, "IBM plans its latest smash hit."

Bacterium Versus Bacterium

Genesis 3:19
"In the sweat of thy face shalt thou eat bread, till thou return unto the ground; for out of it wast thou taken: for dust thou art, and unto dust shalt thou return."

When God finished the creation on the sixth day, He declared all that He made to be "very good." There was no death, no decay, no pain and no sickness. When Adam and Eve committed the first sin, their actions resulted in sickness, decay and death. How could a finished, perfect creation have the ability to become this way?

New research on a strategy to eliminate tooth decay may illustrate an answer to this question. The bacteria most responsible for tooth decay produce lactic acid. Lactic acid eats away at tooth enamel, resulting in tooth decay. Scientists genetically engineered a natural form of this bacterium that produces a chemical that kills the form of the bacterium in our mouths. They then disabled the ability of this bacterium to produce lactic acid. Scientists believe they have developed a harmless bacteria that will remove and replace the harmful bacteria that cause tooth decay. This research raises an interesting question: Is it possible that in the beginning the bacteria in our mouth was harmless but after Adam's fall only a minor modification made it harmful?

No amount of genetic engineering or modern science can reverse the ultimate consequences of sin. Only God's Son, Jesus Christ can take away the guilt of our sin and give us eternal life.

Prayer: Help me, dear Lord, to keep my eyes on heaven where all is perfect. Amen.

Ref: *Science News*, 3/18/00, p. 190, "Wash that mouth out with bacteria!"

Increasing Numbers Want Equal Time for Creation

Proverbs 12:20
"Deceit is in the heart of them that imagine evil: but to the counsellors of peace is joy."

A recent news release revealed the result of a poll on what Americans prefer to have taught in public schools about origins. The poll was done for an organization that favors a dogmatic teaching of evolution. The news release stated only that a huge majority of Americans, 83 percent, favor the teaching of evolution in the public schools. "Creation Moments" investigated this poll more closely, since it's results run contrary to other recent polls.

The professional polling organization polled 1,500 people about whether creation or evolution should be taught in the public schools. Indeed, 83 percent said they believe that the theory of evolution should be taught in the public schools. What was not included in the popular reporting of the poll, however, was that 79 percent of those polled said that creation should also be taught in the schools. Only 20 percent of those polled felt that evolution only should be taught.

Clearly, Americans prefer both creation and evolution presented to the students so that they can make up their own minds. Equally notable is that almost half of those polled agreed with the statement that evolution is "far from being proven to be scientific."

While the statement that 83 percent of those polled favor evolution being taught is true, it leaves out the most important part of the poll's findings. Clearly, ordinary people want to hear about God's creation.

Prayer: Dear Father, help never to be deceived by the forces of darkness. Amen.

Ref: *Christian News*, 4/3/00, p. 2.

Evolution May Be the Victim of Spiders' Webs

Job 8:13-14

"So are the paths of all who forget God; and the hypocrite's hope shall perish: Whose hope shall be cut off, and whose trust shall be a spider's web."

For over a century, scientists have wondered about the embellishments that at least 78 species of spiders weave into their webs. The extra bars and X's seem to have no obvious purpose. Some have suggested that they might be hiding spots, sun shields, bird warnings or insect lures. Now a researcher at the University of Kyoto may have discovered the purpose of these designs. If he is right, evolutionary theory may be the victim of these spiders' webs.

The Asian spider the researcher studied builds two different types of webs. When it is well fed, this spider adds silk bands along the web spokes. However, when this spider is hungry it arranges these bands so that they spiral toward the center of the web. After testing the tension in both web types he discovered that the webs with the spiral bands are much more sensitive to even the smallest insect than the banded webs. In other words, the hungry spider is looking for any insect it might eat. The full spider is only interested in larger insects.

The question is, how could mindless evolution give these spiders the knowledge of structural engineering needed to create these designs? In addition, according to the evolutionary tree for spiders, this knowledge would have to have evolved at least nine times! A much more logical and straightforward explanation is that the Creator gave spiders this knowledge when He made them.

Prayer: Help me, Lord, to do all things excellently, as you have done. Amen.

Ref: *Science News*, 3/25/00, p. 198, "Hungry spiders tune up web jiggliness."

Intelligence Is Built Into the Creation

Psalm 104:24a
"O LORD, how manifold are thy works! In wisdom thou hast made them all."

Those who believe that the universe, the Earth and everything on it are the result of mindless evolution are hard-pressed to explain the origin of intelligence. Some even try to argue that there is no such thing as intelligence, saying that we only think that certain things reflect intelligence. As we learn more about the world, it takes an even greater faith to believe in evolution.

The incredible structure of DNA and how it works is one example. Scientists have now demonstrated, though they did not intend to do so, that DNA is itself infused with intelligence. In 1994, it was discovered that DNA could be used as a computer. Since then, much has been learned about the computing power of DNA.

For example how can DNA be used to solve the problem of finding a flight path between seven cities so that each city is visited only once? To solve the problem, researchers assign a numerical value to individual strands of DNA. Through a series of steps, they allow the strands of DNA to interact with each other, saving the results after each step. An electronic computer can solve a math problem with 30 clauses and fifty variables in 1.6 million steps. A DNA computer can solve the same problem in only 91 steps!

DNA computers would not be possible unless intelligence was built into the very fabric of the creation. Only our all-wise Creator could have been responsible for this.

Prayer: Dear Father, help me use the intelligence you have given me. Amen.

Ref: *Nature*, 1/13/00, pp. 143-144, 175-179, "DNA computing on surfaces."

The Deception of Evolutionary Dating

II Corinthians 11:3

"But I fear, lest by any means, as the serpent beguilded Eve through his subtilty, so your minds should be corrupted from the simplicity that is in Christ."

We have all heard evolutionists claim certain rocks and fossils to be millions of years old. They claim to arrive at these long ages using index fossils. Index fossils are creatures that supposedly lived during specific ages. We are told that these ages are independently verified using radiometric-dating methods. But is all of this as neat and certain as evolutionists say it is?

Recently, a rock formation in England yielded more than just rocks. This rock layer is said to be 189 million years old, based on its index fossils. The index fossils, ammonites and belemnites, were bottom-living deep-sea creatures. Along with these deep-sea creatures, however, partially petrified wood was also found. One piece was found immediately adjacent to a deep-sea fossil. Scientists who believe in young Earth creationism sent samples of the wood to three different radiocarbon dating laboratories. If the index fossil was really as old as evolutionists say, it is almost 4,000 times too old to give a radiocarbon date. Yet all the samples dated at the three laboratories showed consistent young dates. Adjusted by Biblical considerations, they date to the time of Noah's Flood.

We would expect to find land-based life mixed with deep-sea creatures in any rocks that are a result of the violence of the Flood. This find illustrates and supports the biblical account of a worldwide Flood.

> ***Prayer: Lord, the Earth witnesses your Truth. Help me to do the same. Amen.***

Ref: *Creation*, 3-5/00, pp. 44-47, "Geological Conflict."

Rewrite the Evolutionary Textbooks - Again!

I Corinthians 2:13
"Which things also we speak, not in the words which man's wisdom teacheth but which the Holy Spirit teacheth..."

The males of a fish called the three-spined stickleback turn blue-green on the tops of their bodies and red on their bellies when breeding. That is, unless they are male three-spined sticklebacks in the Chehalis river of Washington. These fish turn black when breeding. Evolutionists have long tied this behavior to the fact that the mud minnows in the Chehalis turn black as a threat display. Evolutionary textbooks have long used this difference from other sticklebacks as an example of the evolutionary principle that when two species share a home, they will either become more or less like each other. Now the textbooks will have to be rewritten.

Researchers placed large cages around populations of sticklebacks and mud minnows in the Chehalis River. As they studied the populations, they learned that while sticklebacks will fight with one another, they hardly ever fight with mud minnows. Then, scientists added red and blue-green sticklebacks. Though they showed no black, they had as much success staking out territory as the black sticklebacks. Researchers also found that the black sticklebacks don't turn black until after they stake out their territory. Researchers have had to admit that the differences between the two types of sticklebacks illustrates nothing about how they evolve among different populations.

Unlike Scripture, man's knowledge is always subject to correction. That's why it cannot be used to discredit the Bible, which is knowledge from God.

Prayer: Thank you, Lord, for your truth. Help me grow in your Word. Amen.

Ref: *Science News*, 4/1/00, p. 219, "Fading to black doesn't empower fish."

Talking Caterpillars

Ephesians 4:29
"Let no corrupt communication proceed out of your mouth, but that which is good to the use of edifying, that it may minister grace unto the hearers."

Ants aren't as quiet as you might think, and neither are caterpillars. Through a process called stridulation, both species make sounds. Now, it has been discovered that caterpillars actually communicate with some types of ants using stridulation.

Many species of caterpillars of two butterfly groups secrete a sweet liquid that ants like. In return, the ants protect the caterpillar from its enemies while they collect their sweet treat. These caterpillars actually call the ants when the liquid is ready, using a variety of sounds. These calls range from a simple, "bub . . . bub," to more complex sounds, such as, "beep ah ah ah beep."

Another type of caterpillar calls ants when it needs help. This species does a remarkable job of mimicking the ants' sounds. Very young members of this caterpillar species frequently fall off the leaves on which they are grazing. When they do, they make ant noises until found by foraging ants. The caterpillar's "ant talk" apparently convinces the ants it is one of them. So they return it to their nest and place it in their nursery. There the caterpillar eats some of the young ants. Two ant species eventually figure out their mistake and kill the caterpillar, while a third species never seems to realize its mistake.

Communication is a gift that God has given to many of His creatures. As beings made to have a relationship with God, our communication should be like His, never deceitful and always building others up.

Prayer: Dear Father, teach me how to use words to build others up. Amen.

Ref: *Science News*, 2/5/00, pp. 92-94, "Ants Squeak."

Egyptian Indians?

Genesis 11:7
"Go, let us go down, and there confound their language, that they may not understand one another's speech."

"Creation Moments has investigated a number of New World sites at which ancient Old World writing has been found. These examples, of which there are many, serve to show that before the Dark Ages, knowledge of the New World was fairly common and that man has always been curious and resourceful.

One of the most unusual examples of ancient Old World inscriptions in the New World is found in northwestern Oklahoma. The inscriptions are located on the walls of Anubis cave. If you think that name is Egyptian, you are correct. Inside the cave are a number of ancient Egyptian hieroglyphs. These include a human figure standing on a three-dimensional cube. Rays extend from the figure's head. Anubis, the Egyptian god that was said to conduct the souls of the dead to judgment, also appears. The oddest element among the figures, however, is the presence of Celtic ogam inscriptions. These inscriptions indicate that the cave was used for Celtic ceremonies. Were the Celts familiar with Egyptian hieroglyphs? Or was this North American cave visited in ancient times by both Egyptians and Celts?

It was not until God confused man's language at the Tower of Babel that man took seriously God's command to spread out and fill the Earth. With their new languages, people were more motivated to use their abilities to explore the entire world. Ancient Old World inscriptions in the New World illustrate that these God-given abilities have always been considerable.

Prayer: Lord, thank you for the beauty of world you have given us to explore. Amen.

Ref: *Science Frontiers*, p. 33, "Did the Ancient Egyptians Explore Oklahoma?"

The Tree With Nothing to Offer But Rot

Genesis 11:1b

"And God said, 'Let the earth bring forth . . . and the fruit tree yielding fruit after its kind, whose seed is in itself, on the earth'; and it was so."

The chempedak tree is found in Malaysia and is closely related to jackfruits in the mulberry family. The chempedak tree produces edible fruits that are about a foot long.

Researchers studying the chempedak were surprised to discover that its blooms irregularly. One tree was watched by researchers for five years before it bloomed! While the flowers smell like watermelon, they offer no nectar to entice pollinating insects. Moreover, the insects that do pollinate the chempedak tree do not eat any of the pollen. What, then, attracts the two species of gall midges that pollinate the flowers?

The answer: a fungus that attacks only the male flowers of the tree. The midges like to eat the fungus, so they crawl on the male flowers as they eat the fungus, picking up pollen. Then they go to the clumps of female flowers looking for more fungus, inadvertently pollinating them.

This arrangement, in which the tree depends entirely on a fungus to attract pollinators, is unknown in any other species. This raises the question of how such an arrangement could have evolved. Without both the midges and the fungus, the first chempedak tree would have been the last. All three elements in this interrelated system had to have come into being at the same time.

The chempedak tree bears witness to a rapid creation of the interdependent plant and insect world where living things had to wait, at most, mere days for the other creatures it depends on to come into existence, just as Genesis reports.

Prayer: Dear Father, with the creation I will bear witness that you are Creator. Amen.

Ref: *Science News*, 3/18/00, p. 182, "Tree pollination needs male-only rot."

Do Birds Truly Make Music?

Song of Solomon 2:12
"The flowers appear on the earth; the time of singing of the birds is come, and the voice of the turtle is heard in our land."

The Bible speaks of bird calls as songs, as most of us do. But, evolutionary theory has led some scientists to say that we are merely assigning human meanings to the calls of birds. They say that the bird calls have nothing to do with real music.

Ornithologists have known for some time that bird songs use the same musical scales as our music. Decades ago it was noted that some of Beethoven's work could be heard from the European blackbird. The music was the same as the opening rondo of Beethoven's *"Violin Concert in D, Opus 61."* Since these birds pass their songs from generation to generation, Beethoven could have gotten the lilting music from the forefathers of today's European blackbird!

The songs of some species, like the song sparrow, follow the form of a sonata. The songs begin with a strong theme, then the theme is playfully manipulated, and for a finish, the original theme is repeated. Mozart had a starling as a pet. Once, having heard Mozart play his *"Piano Concerto in C Major,"* the starling not only imitated it, but also changed the sharps to flats! Mozart exclaimed, "That was beautiful!" When the starling died, Mozart held an elaborate funeral for it. Eight days later he wrote, *"A Musical Joke,"* which contains the same elaborate structure found in starling song.

Do birds make true music, as the Bible says? Contrary to what some evolutionists say, Beethoven and Mozart certainly thought they did.

Prayer: Lord, I thank you for the gift of music and I await the music of heaven. Amen.

Ref: *Science News*, 4/15/00, pp. 252-254, "Music without Borders."

Those Clever Whales, Dolphins and Seals

Psalm 148:7

"Praise the LORD from the earth, ye dragons, and all deeps..."

Scientists have long wondered how deep-diving creatures such as the whale and the dolphin can dive so deeply on one breath. It takes a great deal of energy for a seal, for example, to dive 1200 feet deep and then return to the surface. Scientists computed that one breath should not provide enough oxygen to burn the muscular fuel necessary for such dives.

In one attempt to learn how deep-diving sea creatures do what seems to be impossible, researchers even built a robotic fish. Their studies of how the fish swam indicated that gliding would not conserve a deep-diving creature's energy. Now, scientists have finally been able to attach video cameras to deep divers and have discovered secrets of deep-diving animals. Researchers found that these animals actually do save energy for their return to the surface by gliding into the depths. To their surprise, researchers saw that as the animals descend, water pressure squeezes their bodies into a smaller volume, making them more dense. The pressure also flattens air sacs in their lungs, further helping the animal to descend. These effects help save oxygen reserves for hunting and the return to the surface.

No one has even seen cleverness result from the chance forces that supposedly drive evolution. Intelligent design comes from an intelligent Designer, our Creator God.

Prayer: Dear Father, help me to glorify you with my actions. Amen.

Ref: *Science News*, 4/8/00, p. 230, "How whales, dolphins, seals dive so deep."

223

Lima Bean Self Defense

Psalm 20:1
"The LORD hear thee in the day of trouble; the name of the God of Jacob defend thee..."

Several plant species are equipped to poison or repel a predator. The lima bean, however, employs a more sophisticated defense strategy that not only warns surrounding lima beans of the attack, but also calls in defenders.

Anyone who has ever worked with plants is familiar with spider mites. There are many varieties, but one of the most dangerous to the lima bean is the two-spotted spider mite. These mites inject their saliva into a plant's tissues, dissolving them. One of the mites' biggest enemies is a carnivorous mite that feeds on these two-spotted mites. These carnivorous mites are small, and travel wherever the winds carry them.

When two-spotted mites attack a lima bean the plant sends out a special chemical signal. When this distress signal reaches surrounding lima beans they too begin to send the signal, even though they are not under attack. The signal carries several messages to different species. Spider mites not on the plant being attacked are repelled from the area. On the other hand, any carnivorous mites that drift into the area will stay to feed on the two-spotted spider mites. As a result, the lima bean defends itself by sending specific messages to three different species!

The lima bean's intelligent defense strategy is difficult for evolutionists to explain. But is easily accounted for by a Creator Who cares about all His creatures.

Prayer: Lord, thank you for defending me against sin, death and the devil. Amen.

Ref: *Bombardier Beetles and Fever Trees*, pp. 28-20.

Study Shows Americans Are Spiritually Hungry

2 Timothy 1:13
"Hold fast the form of sound words which thou hast heard of me, in faith and love which is in Christ Jesus."

A new poll conducted by George Gallup reveals some alarming trends in Americans' religious thinking. The good news is that more than 80 percent of Americans desire to grow spiritually. And Christian religion continues to show wide popularity, with church attendance remaining steady over the last half century. Also positive is that people are seeking a more meaningful and deeper spiritual life. Gallup attributes this to the failure of materialism to satisfy the human heart.

The bad news is that Gallup's study found that there is widespread ignorance about the Bible, basic Christian teachings and traditions. Even more disturbing is that the study found that Christians generally don't know what they believe or why. As a result, Gallup found a widespread tendency to mix Christian and non-Christian beliefs. People tend to pick and choose what they want of Christianity, along with beliefs from other religions. For example, Gallup cites evangelical Christians who also believe in reincarnation. Here at Creation Moments, we frequently run into people who profess to be Bible-believing Christians, and still believe that God used evolution to make the world we know today.

A good part of our ministry is to help people understand what the Bible says and to know why they believe it is true. But this job belongs to every Christian. When each of us takes this seriously, we will be able to offer others the true spirituality that so many are now seeking.

Prayer: Dear Father, grant me a more complete understanding of your Word. Amen.

Ref: *Reporter*, 4/00, p. 8a, "Gallup: Americans embrace `pick-and-choose' faith."

Extra-biblical Writings Support the Bible's History

Genesis 11:8
"So the LORD scattered them abroad from thence upon the face of all the earth: and they left off to build the city."

Many people today tend to dismiss the early chapters of Genesis as legend rather than history. But, if the Bible's story of the Flood and the Tower of Babel are not history, there are some established historical facts that have no explanation.

According to the Bible, after the great Flood, people multiplied but failed to scatter over the face of the Earth as commanded by God. Having one language, most of the people stayed in one location and built a great city. It was not until God confused their language that they scattered across the Earth. If these events did not take place, how do we explain the fact that a 6th century B.C. Chinese writer describes the ultimate God Who created all things in much the same way as an ancient Egyptian writer? And why would these descriptions match 8th century-B.C. Greek writings?

While it is true that each of these cultures had their unique gods, their ancient writings about the Creator use very special language. Lao-tzu was an 8th century-B.C. writer who described the ultimate God as eternal and self-existing. Likewise, an ancient Egyptian writer described the ultimate God as the source of everything in the heavens and on Earth as well as all life.

If man was not once of one culture and language, how did all these ancient writers each get the same story to tell? The similarities between these descriptions and those of other cultures bear witness to the historicity of even the earliest parts of Genesis.

Prayer: Thank you, Lord, for your saving Word that tells of my salvation. Amen.

Ref: *After the Flood*, pp. 16-19.

Can Man Become Immortal?

Psalm 23:4
"Yea, though I walk through the valley of the shadow of death, I will fear no evil; for thou art with me; thy rod and thy staff they comfort me."

When Adam and Eve sinned in the Garden of Eden, they brought death into the world. Death is an enemy, but for the believer who has eternal life, death provides a doorway out of a sinful world and into Christ's presence. Modern science has learned that our cells themselves carry the information that will result in their death one day. Add to this the wear and tear of everyday life, and accidents, and death becomes inevitable.

Some modern scientists' however, believe that they are within reach of doubling our life spans. Some even think they can do away with death completely. Each time a cell divides, the tips of the DNA become shorter. When the DNA chains shorten to a preprogrammed length, the cell dies instead of dividing again. Researchers have now discovered a protein that can restore those tips, theoretically preventing a timed death. Scientists have also learned how to inject new neurons into the brain to restore it to youth. They are also experimenting with methods of growing new organs for transplantation. These and other procedures, they say, could end death forever. Unfortunately, some of these advances come at the cost of the lives of aborted babies and lab grown-human embryos.

Can man reverse the death that has resulted from his own sin? Only one Man can. Jesus Christ, the man-God, gives eternal life to all who believe in Him as their Lord and Savior and makes death but a shadow for the believer.

Prayer: Thank you, Lord, for giving me the forgiveness of sins and eternal life. Amen.

Ref: *U.S. News & World Report*, 3/20/00, pp. 58-59, "The cells of immortality."

An Ancient Sweet Solution to a Sticky Modern Problem

Psalm 81:16
"He would have fed them also with the finest of wheat: and with honey out of the rock should I have satisfied thee."

We have all heard of "superbugs"—bacteria that are resistant to antibiotics. Antibiotic resistant Staphylococcus bacteria have even shut down hospital wards. Now researchers in New Zealand and Australia may have found out how to fight superbugs. And their weapon is one that has been in medical use for over 4,000 years!

Researchers in New Zealand have examined the healing properties of over 30 different types of honey native to New Zealand. They have found that honey made from the manuka bush has some powerful healing properties. As the honey mixes with the fluids in a wound it produces a small but continuous stream of hydrogen peroxide. Being acidic, honey creates a hostile environment for bacteria. Research is being conducted using honey to treat burns, eye infections, diabetic skin ulcers and other conditions with encouraging results. Australian researchers have found that honey will even kill 100 different strains of antibiotic resistant bacteria including the flesh-eating bacteria! One New Zealand man who had developed gangrene that could not be controlled with antibiotics tried honey. The honey stopped the infection and his foot began healing. (Please note that the honey available in stores has been heated or pasteurized, removing its antibiotic properties).

Honey is a gift of God that has many uses. That it has been used on wounds since earliest times suggests that God gave the first people knowledge of its healing powers.

Prayer: Dear Father, please keep me in good spiritual and physical health. Amen.

Ref: *Rx Remedy online*, "The Healing Powers of Honey."

Evolutionary Scientists Find a Surprise!

Matthew 24:44
"Therefore be ye also ready: for in such an hour as ye think not the Son of Man cometh."

According to the Bible's history, human beings were created perfectly by God about 6,000 years ago. We should note that this means we were not only morally perfect, but genetically perfect as well. According to evolutionists, our ancestors split off from the monkeys millions of years ago. Both viewpoints agree that each generation of human beings adds mutations to the ongoing human gene pool.

How many mutations are added by each human generation? Researchers with the University of Sussex and the University of Edinburgh looked for the answer to this question. They examined DNA from living parents and children, as well as known mutation rates for mammals. They assumed that some mutations would be fatal and thus not added to the overall gene pool. They also assumed that a very few mutations might be helpful, an assumption that has never be demonstrated to be true. They finally calculated that 100 harmful mutations are added to the human gene pool by each generation. Their published conclusion was that, considering the millions of years humans have been evolving, we should have easily accumulated enough mutations to make us extinct a long time ago!

This research is clear evidence for a recent appearance of man, and implies that the Earth was created a relatively short time ago, as portrayed by the Genesis. However, human life on Earth will end when Jesus Christ returns to collect all believers to live in a new heavens and Earth.

Prayer: Lord, your Word is true so I eagerly await your return for me. Amen.

Ref: *Nature*, 1/28/99, "High genomic deleterious mutation rates in hominids," pp. 344-347.

Crooked Critters

Colossians 2:8
"Beware lest any man spoil you through philosophy and vain deceit, after the tradition of men..."

Con men develop elaborate schemes to cheat people out of their money or other possessions. Successful cons are cleverly and intelligently directed toward the conman's goal. We would not expect such intelligently directed, albeit sinful, behavior in animals, except that the Bible tells us that deceit is the way of the world since sin has entered the creation.

To their surprise, two groups of researchers have discovered a number of animals that try to con their own kind. One group of researchers spent two years standing in ponds, getting to know green frogs. They discovered that when an intruding frog enters a frog community, even the younger males attempt to intimidate him by croaking at a much lower pitch than normal. This makes even the young frog sound like the bigger members of the community, bluffing the intruder. Another team studied Afro-Asian fiddler crabs. When one of the male crabs loses its claw, it will grow back only slowly. So the crab quickly grows a fake claw until the new one grows in. The fake claw is useless for fighting, but works well in aggressive claw waving at other males. It can also be used in the claw waving that is part of the crabs' mating ritual. Females prefer the fastest waving claws, so the lighter fake claws, which wave faster, are a real advantage.

While animals cannot be charged with sin, we are responsible for our actions. We thank God that he has sent His Son to save us from the guilt our sins.

Prayer: Dear Lord, show me where I have been fooled into misbelief. Amen.

Ref: *Science News*, 4/22/00, p. 262, "The truth is, frogs bluff and crabs cheat."

The Truth About Neanderthal DNA

Acts 17:26
"And hath made of one blood all nations of men for to dwell on all the face of the earth, and hath determined the times before appointed, and the bounds of their habitation..."

In 1997 scientists recovered mitochondrial DNA from the leg bone of a Neanderthal man. After comparing it to modern human mitochondrial DNA, they declared that Neanderthal man was not human after all. This neatly moved him back into the ape-man category, but we need to look at this claim more closely.

Neanderthal skeletons have the same number of bones as ours. It is true that their bones were thicker and stronger than ours. Their bodies were also stockier and, typically, their arms and legs were shorter in proportion to the rest of the body. These differences, however, are also found in modern humans who live in cold climates, as many Neanderthals did. In a cold climate a stockier body and shorter limbs reduce body heat loss. Neanderthals who lived in warmer climates had longer arms and legs.

We know of at least 36 instances where Neanderthals buried their dead, often with flowers and other items suggesting religious ceremony. One Neanderthal grave also contained a small bone flute. No animal makes tools to use in making more complex tools, but Neanderthal did. The simple fact is that the Neanderthal DNA is at the end of the modern human range.

All human beings, including Neanderthal, were created by God. What separates us from animals is that God sent His only Son to die for us so that, believing in His salvation, we may have eternal life.

Prayer: Help me, dear Lord, to love all people that you made as you do. Amen.

Ref: *Impact (ICR)*, 5/00, "Neanderthals Are Still Human!"

How Fast Do Stalactites Grow?

Genesis 7:11a
"In the six hundredth year of Noah's life, in the second month, the seventeenth day of the month, the same day were all the fountains of the great deep broken up..."

For generations cave guides have been telling cave visitors that the stalagmites and stalactites that they see around them grow at incredibly low rates. The figure that is usually given is one inch per century for stalactites (they are the ones that hang down). That estimate, however, was not based on any scientific research. Even in the 1970s, the growth of these formations was not fully understood.

Now a group of scientists who believe in young Earth creationism has studied the chemistry of stalagmite and stalactite formation in actual cave conditions. They discovered that there are many variables in the growth of these formations. How much carbon dioxide dissolves in the water depends on temperature and pressure. The amount of carbon dioxide in the water can vary by a factor of five. The dissolved carbon dioxide turns the water acidic, causing it to dissolve calcium carbonate, that is, limestone. More turbulent water can dissolve more calcium carbonate. The same water under pressure can dissolve even more. Once the water seeps out of the cave ceiling, the pressure is relieved, the water evaporates, and the calcium carbonate begins to deposit to form the stalactite. Actual growth rates in actual caves have been observed as high as one inch every seven and a half days!

The violence and high pressures developed during the Genesis Flood would have provided the ideal conditions for producing a great deal of underground water supersaturated with calcium carbonate.

Prayer: Dear Father, prosper true and right knowledge among us. Amen.

Ref: *CRSQ*, 3/00, pp. 208-214, "What is the Upward Limit for the Rate of Speleothem Formation?"

Survival of the Most Cooperative

Hebrews 13:6
"So that we may boldly say: 'The LORD is my helper; and I will not fear what man shall do unto me?"

"Survival of the fittest" is an essential principle of evolution. This principle has not only been applied to animals, but also to human beings, as a social theory.

Scientists, however, are learning that cooperation among animals is more often the rule. Researchers studied seven pairs of unrelated capuchin monkeys in the lab to see if they would cooperate. The monkey pairs were then placed in an enclosure, one on each side of a mesh screen. Two clear bowls of apple slices were placed on trays designed to be accessible to the monkeys when they pulled on a bar. However, the acquired food would be given only to one of the monkeys. When the bar was set so that both monkeys had to pull on it to get the food, the monkey that received the food shared with his helper. When the bar was set so that one monkey could access the apple slices, he generally didn't share. Clearly, the monkey who came into control of the food with help, felt that it was his duty to reward his helper. Instead of talking about "survival of the fittest," one researcher spoke about the "deep evolutionary roots of cooperation". Unfortunately for evolutionary scientists, real science very often gives good evidence that contradicts the theory of evolution, but we seldom hear of this.

God has designed the creation so that we sometimes need help, or have the opportunity to help others. He did this to remind us that He is our ultimate Helper.

Prayer: Dear Father, I trust you to help me even when I don't know I need help. Amen.

Ref: *Science News*, 4/8/00, p.231, "Cooperative strangers turn a mutual profit."

There Is No Genetic Code for the Human Soul

James 2:26
"For as the body without the spirit is dead, so faith without works is dead also."

The $250 million Human Genome Project is attempting to map all the human genes. Some observers have been saying that the human genome project will define what it means to be human.

Such talk, however, is based on evolution. We are not the sum of our DNA. It is well known, for example, that two people can inherit the same weakness for alcohol. One may become an alcoholic, while the other avoids alcohol for his entire life. This does not mean that mapping the human genetic code is inherently evil. Such knowledge can lead to precisely designed drugs that do what they are supposed to do without any side effects. It may also lead to better treatments for cancer, Alzheimer's and other diseases that can have a genetic cause.

At the same time, dangers are beginning to appear. For example, with the knowledge of which particular genetic information produces what disease, unborn babies could be screened and those who have a potential to develop unwanted health problems might be aborted. A British governmental agency has already recommended this course of action. Another danger is that children may be forced into certain careers simply based on their "genetic potential." One's genetic profile may even lead insurance companies to refuse coverage to people whose genes suggest the possibility of some future disease.

All our genetic information added together does not define who we are because we also have a soul that has no genetic code and only the knowledge of Jesus Christ can bring health to the human soul.

Prayer: I thank you, Lord, for bringing health to my soul through forgiveness. Amen.

Ref: *World*, 4/29/00, pp. 18-21, "Cracking the Code."

Were Dinosaurs Warm-Blooded?

Psalm 147:2-3
"The LORD doth build up Jerusalem: he gathereth together the outcasts of Israel. He healeth the broken in heart, and bindeth up their wounds."

Dinosaur fossil hunters and paleontologists have long debated whether dinosaurs were warm-blooded or cold-blooded. With reptile characteristics, it was expected that dinosaurs would be cold-blooded. Yet, some evidence indicated that they must have been warm-blooded. The problem is that their soft organs are almost never fossilized.

The words "almost never" are important. A dinosaur liver and diaphragm—the soft parts—were recently found fossilized. Fossilization of soft organs is a testimony to the rapidity with which the dinosaurs were encased in the mud that later hardened to rock. Now a fossilized dinosaur heart has been discovered in South Dakota within the chest cavity of a fossilized Thescelosaurus.

Once researchers recognized that the stone within the chest cavity might be a heart, they X-rayed it. X-rays revealed a four-chambered heart with one aorta. Reptiles have two aortas. In other words, *Thescelosaurus* had a heart constructed like the hearts of warm-blooded animals. This kind of heart is more efficient in delivering oxygen to the body. This in turn supports a more active lifestyle and higher metabolic rates. Based on the structure of *Thescelosaurus's* heart, many scientists have concluded that it was indeed warm-blooded.

The same Creator who made the dinosaur's heart is able to heal our hearts when they are broken by sorrow. It is He Who heals our broken hearts through the forgiveness of sins that is found in Jesus Christ.

Prayer: Thank you, Dear Father, for bringing your power into my life. Amen.

Ref: *Science News*, 4/22/00, p. 260, "Telltale Dinosaurs Heart Hints at Warm Blood."

Ginger Is Strong Medicine

Ezekiel 47:12b

"...it shall bring forth new fruit according to his months, because their waters they issued out of the sanctuary: and the fruit thereof shall be for meat, and the leaf thereof for medicine."

Naturally occurring medicines often work better and more gently than man-made concoctions. Ginger has long been accepted as a natural treatment for nausea. Several modern studies have confirmed that ginger can even work better than modern nausea medicines such as Dramamine®.

The ginger in question comes from the roots of tropical plants that grow two to four feet high. It was once used to calm bees so that honey could be harvested without incident. A 1988 study gave either ginger capsules or placebos to Dutch naval cadets. The study concluded that ginger proved to be "significantly beneficial" in reducing the incidence of seasickness. Two studies in the 1990s showed that ginger reduced nausea and vomiting associated with chemotherapy and some types of surgery. A Utah study compared the effectiveness of ginger versus Dramamine®. Students were rotated on a tilted chair for six minutes. None of those who had taken the Dramamine® made it through the six minutes. Half of the students who had taken ginger lasted the full six minutes. (Researchers warn that you should check with your doctor before trying ginger since it does interact with some medications.)

When God placed helpful natural medicines into His perfect creation, He was providing healing for us before we even needed it. This care shows that we have a loving Creator. His greatest gift to us is the spiritual healing of the forgiveness of sins that comes through His Son, Jesus Christ.

Prayer: Lord, I thank you for all your goodness to me both now and forever. Amen.

Ref: *Rx Remedy*, 1999 (on-line), "Avoiding Nausea - Gingerly."

How the Aye-Aye Taps Into Lunch

Isaiah 45:18
*"For thus saith the LORD that created the heavens; God himself
that formed the earth and made it; he hath established it, he
created it not in vain, he formed it to be inhabited: 'I am the LORD,
and there is none else.'"*

The Aye-Aye is one of the strangest little monkeys on Earth. Its
peculiar features bear witness that it was specially designed and created to
fill a unique niche in nature, not a chance development of evolution.

The Aye-Aye lives in Madagascar. This small creature has huge
ears and an extra long middle finger. It crawls on trees, tapping on them to
find grubs. The Aye-Aye's ears are so sensitive that it can tell if there is a
grub hole hidden beneath the surface of the wood. Not only can the
Aye-Aye sense a grub hole an inch within the wood, it can even tell if there
is a grub on the hole. When the Aye-Aye senses a grub, it chews into the
wood with forward curving, incisor teeth. However, the Aye-Aye doesn't
have to worry about wearing its teeth down chewing on the tough wood. Its
special incisor teeth are unique among primates. They continue to grow
throughout the Aye-Aye's life, just like a rodent's teeth. Once down to the
grub, it uses its elongated middle finger to fish the grub out for lunch.
Researchers have noted that the Aye-Aye fills the same niche as the
woodpecker in an environment where there are no woodpeckers.

The Aye-Aye has clearly been specially designed with unique
features so that it can make its living in a niche that is unfilled in its own
environment. Our Creator God not only made the Earth to be inhabited, He
designed His creatures to make a living within it.

> ***Prayer: I thank you, Dear Father, for the unique richness of
> your creation. Amen.***

Ref: *Science Frontiers*, pp. 136-137, "The Aye-Aye, A Percussive Forager."

Who Was First to North America?

Genesis 1:28
"Then God blessed them, and God said unto them, 'Be fruitful and multiply; and replenish the earth...'"

The conventional understanding has always been that North America was populated by people who crossed from Asia to Alaska via the Bering Strait and then migrated down the West Coast. Eventually, some of those people crossed as far as the eastern coasts of North and South America. Known as the Clovis culture, their descendants produced all the North and South American Indian tribes and nations. The Clovis sites that have been excavated so far have produced distinctive spear points that have become known as Clovis points.

That understanding has now been challenged by finds in Virginia. Signs of human habitation have been found at Cactus Hill, which is 45 miles south of Richmond. Researchers have dated the site to be fifty percent older than any known Clovis excavation. Stone points and blades have been found at Cactus Hill which show wear marks typical of hide scraping and butchering. The points and blades are unlike any known Clovis points; however, they do resemble western European blades and points from the same period. These facts have led some researchers to conclude that the site is evidence that ancient Europeans crossed the Atlantic Ocean to begin a settlement in North America. This was long, long before the time of the Vikings or Columbus.

When God made Adam and Eve, and again, after the Genesis Flood, God commanded human beings to "fill the Earth." More and more we find evidences that early man obeyed this commandment.

Prayer: Dear Father, help me to be a good steward of all you have given me. Amen.

Ref: *Science News*, 4/15/00, p. 244, "Early New World Settlers Rise in East."

The Stranger Who Isn't a Stranger

Genesis 17:8

"And I will give unto thee; and unto thy seed after thee, the land wherein thou art a stranger, all the land of Canaan, for an everlasting covenant, and I will be their God."

Social insects seldom, if ever, tolerate other species within their communities. Strangers are generally identified by their smell, rather than by their appearance, since those insects that aren't actually blind have very poor eyesight. It is common for social insects to unite to destroy even one stranger who invades their nest.

Like most social insects, termites don't tolerate strangers. However, the *Trichopsenius* frosti beetle likes to make its living within the nest of just one species of termite. These termites produce a very specific mixture of 21 compounds that are mixed into a unique scent and used to identify nest mates. The unique scent includes a compound that is a third of the total mix, while other compounds are included as less than one percent of the mix. As unlikely as it sounds, the *Trichopsenius* is able to make the same mix of compounds so perfectly that its presence in the termite nest goes unquestioned.

Those who trust Christ for their salvation are strangers in this world. But unlike the *Trichopsenius* beetle, because we are strangers here, we should not strive to be like those among whom we are strangers. Like Abraham, those who trust in Christ have been given God's promise that He will lead them through this wilderness to a new and perfect land that will last forever. Abraham could not guess how God would make his descendants a great nation and give them a beautiful land. Nevertheless, he trusted God's promise, and we can, too!

Prayer: Dear Lord, teach me how to be like you and live as your child. Amen.

Ref: *Bombardier Beetles and Fever Trees*, p. 46.

Perception Can Be Relative

Isaiah 6:9

"And he said, 'Go, and tell this people, Hear ye indeed, but understand not; and see ye indeed, but perceive not.'"

A primary step in understanding how we think is realizing that our perception of the world around us is based upon unprovable beliefs. Those who do not believe in God will see no evidence of His working in the world.

In November of 1919, the name Albert Einstein became a household word. That was when Arthur Eddington announced that Einstein's theory of relativity had been confirmed as "one of the highest achievements of human thought." This was based on one photograph, taken during an eclipse of the sun, in which starlight was said to be bent by the sun's gravity as much as Einstein had predicted.

Modern researchers have been reviewing the evidence. Some photographs were taken by teams in Brazil and in Africa. Modern researchers have discovered that of all the photographs taken, only one shows the bending Einstein predicted. Most of the plates show bending on the order predicted by Newton. Eddington simply ignored all those other plates. To make matters worse, any evident bending of light was so small that the same effect could have been produced by distortion by the Earth's atmosphere or the changing temperatures in the telescope. In short, Eddington perceived only what he wanted to perceive!

The same is true of those supposed evidences for evolution. Not one example of a cited evidence for evolution is more than a perception based upon the assumption that evolution actually occurred. This is definitely not the objective method that science claims.

Prayer: Dear Father, help me to perceive the world in terms of your love. Amen.

Ref: *Journal of Scientific Exploration*, Vol. 13:2, p. 271 ff., 1999.

Vision Reveals God's Hand

Proverbs 20:12
"The hearing ear and the seeing eye, the LORD hath made even both of them."

Did you know that if you stared at an object with absolutely no eye movement you would eventually go blind? Actually, it is impossible to stare at anything without some eye movement, and there is a good reason for this.

You may be aware that each eyeball has a blind spot where the optic nerve is attached. Normally we don't notice it because the other eye fills in the missing portion of the image in our brain. But, there is much more than this going on to make vision possible. Suppose that you are talking with someone, nodding your head in agreement. Despite the movement, your eyes stay focused on his face. You can do that because of an organ in your ear called the vestibular sensory organ. It senses the motion of your nodding head, and sends the information to your eyes, enabling them to stay focused while your head moves.

More amazing is what you don't see. Your retina is crisscrossed with blood vessels that cast shadows into your field of vision. You just don't see them because the shadows never move. This explains why, if you were able to stare perfectly at something, you would go blind. Your brain automatically filters out perfectly nonmoving images. If it didn't, you would constantly see the web of blood vessels on your retina. So when you stare intently at something, the muscles of the eye send microscopic tremors to the eyeball making it vibrate so that you won't go blind.

The ability to see is the result of several precision-designed systems all working together to give us reliable vision. God, not chance, creates such precise integration.

Prayer: I rejoice, Lord, for you have not left my design to chance. Amen.

Ref: *Discover*, 6/00, p. 108, "Your Steadicam."

How Important was Jerusalem in David's Day?

2 Samuel 5:7
"Nevertheless David took the stronghold of Zion: the same is the city of David."

The Bible portrays Jerusalem as an established city of some importance even at the time of Abraham. Yet many Biblical scholars describe Jerusalem as, at best, an unimportant settlement as late as the time of David in 1000 B.C. Now a biblical scholar may have found the evidence to settle the issue in favor of the biblical history.

The new finds were all previously documented, but they were considered only curiosities. The first piece of evidence was found underneath the site of a Byzantine church in Jerusalem. It is a slab of stone mounted as a monument. Both sides of it are covered with Egyptian hieroglyphics. Another team of archaeologists discovered two Egyptian alabaster vessels in Jerusalem. Their style dates them to the years of 1575 to 1308 B.C. Also beneath the church was a marble-like slab of stone with channels cut in it. While nothing like it has ever been associated with a Byzantine church, such slabs are commonly found as offering tables in Egyptian temples. Now an excavation in Jerusalem has actually uncovered the site of an Egyptian temple and, finally, Egyptian records dating to the 14th century B.C. contain six letters from the Egyptian provincial ruler of Jerusalem.

Added together, these evidences not only describe Jerusalem as a city, but one of some importance, just as the Bible describes. It was important enough for the Egyptians to build a temple to one of their gods in Jerusalem itself and for David to make it his capital.

> ***Prayer: I thank you, Lord, that even the history in the Bible is reliable. Amen.***

Ref: *Biblical Archaeology Review*, 5-6/00, pp. 48-57, 67, "What's an Egyptian Temple Doing in Jerusalem?"

Does Our Blood Prove Evolution?

Ephesians 2:13
"But now in Christ Jesus ye who sometimes were far off are made nigh by the blood of Christ."

One of the early evidences for evolution, at one time found in almost every school textbook, was based upon the salt content of human blood plasma and the salt content of the sea. Professor Macallum of the University of Toronto began with the assumption that life on earth began in the sea and that the first land creatures had retained that sea salt concentration in their blood. This assumption was then turned and used as textbook evidence that evolution had, in fact, taken place.

Such so-called evidence may appear laughable today but we should be reminded that what was in school textbooks had to be learned and is still in the collective consciousness of many today. Just what are the facts? Human blood plasma has 250 times more iron and 9,000 times more selenium than seawater. On the other hand, seawater contains much more magnesium than our blood, so it is simply untrue that the mineral content of our blood reflects that of seawater of long ago.

Blood plasma does contain between 20 and 30 percent of the amount of salt in present day seawater. While this may appear to be a good argument for evolution, it contains a fatal flaw. According to evolution theory, the first land animals emerged from the sea 350 million years ago, but from the known rates at which minerals are being added to the sea, the salt content at that time would have been virtually zero and the mineral content even less!

Our blood offers no proof of evolution. But the shed blood and Resurrection of Jesus Christ proves His victory over sin, death and the devil.

Prayer: I thank you, Lord, for your sacrifice so I might live eternally Amen.

Ref: *Creation*, 3-5/97, "Red-blooded Evidence."

When a Female Bee Isn't Even a Bee

2 Corinthians 1:11
"...ye also helping together by prayer for us, that for the gifts bestowed upon us by the means of many persons thanks may be given by many on our behalf."

Blister beetles of California's Mojave Desert depend on solitary bees for their life cycle. The beetles, however, have nothing of interest to offer the bees.

Blister beetle larvae are so tiny that dozens of them can infest a solitary bee's body. Riding on the female bee, they transfer into the solitary bee's nursery when the female lays her eggs. There the beetle larvae eat the pollen that the mother has packed there for her hatchlings. Once they pupate into wingless adults, the blister beetles then need a male bee to carry them to a female so the next cycle of life can begin. To attract a male bee, large numbers of the beetles pile together into a clump that looks like a female bee. They will hold this shape for up to 2 weeks, waiting for a male bee to show interest. Researchers have also concluded that while in this position the beetles also generate the scent of a female bee ready to mate! Once a male bee gets close enough, the tiny beetles jump on his body. When he mates with a female, the beetles transfer to her body and wait for her to lay eggs. Scientists are amazed that the beetles, which are not social insects, are smart enough to work together to fool the male bees.

Obviously the beetles did not design this clever strategy by themselves. The cooperation they show for their survival was designed and programmed into them by their wise Creator, perhaps to show us how important working together is for survival.

Prayer: Lord, help your people work together for the good of your kingdom. Amen.

Ref: *Science News*, 5/6/00, p. 295, "Ah, my pretty, you're...#&! a beetle pile!"

"Darwin's Finches" - No Proof of Evolution

Genesis 1:21

"And God created . . . every winged fowls after his kind: and God saw that it was good."

During his visit to the Galapagos Islands, Charles Darwin saw that each island was populated by a little finch. These birds were all very similar, yet from one island to another there were some differences in the size and shape of the beak. Darwin reasonably concluded that in the distant past a pair of these birds had been blown 600 miles from the mainland and had since multiplied and spread. The environment on each island was slightly different and the birds had developed specialized beaks to exploit the different food sources; it seemed to Darwin that here was a unique example of evolution in action where 13 different species had arisen from just one mating pair.

A species is defined by the ability to reproduce, thus a sterility barrier separates one species from another. Textbooks use Darwin's finches to claim that new species have been produced, thus demonstrating evolution in action. The facts are that at least six of these different birds are known to interbreed and thus by definition are not new species at all but simply varieties within a single species. Moreover, DNA studies reveal very little difference between any of these birds and there is no evidence of new genetic material, which is essential if evolution actually took place.

Darwin's finches offer no support for evolution. They do support the Bible when it teaches that, like every other creature, the birds have simply reproduced "after their own kind." God has so designed His Creation that while it is possible for creature to adapt perfectly to individual environments, there is still stability of the basic "kind."

Prayer: Dear Father, I thank you for your Word that tells me of salvation. Amen.

Ref: *Creation Matters*, (CRS), pp. 5-6, "Quiz."

Don't Eat the Red Snow

Psalm 147:16-17

"He giveth snow like wool: he scattereth the hoarfrost like ashes. He casteth forth his ice like morsels: who can stand before his cold?"

Over 2,000 years ago Aristotle wrote about red snow. Today red, orange and even green snow has been found on every continent. It is typically found in deep mountain snows in the late spring. The snow color is caused by any of 350 species of strange snow algae.

The pigments inside the resting algae cysts produce the colors. Even while resting in the nutrient-poor, highly acidic snow, the cysts are photosynthesizing. In its cyst form, the algae can live even in temperatures of -94°F. As the snow melts in the late mountain spring, the cysts burst open. A single cell with two whip-like tails emerges. It then begins to swim upward, against the draining water from the snow above. At this time of the year, the snow melt water can reach the same acidity as a peat bog, which causes death to most microbes. Those that manage to swim through the snow mate and then return to the cyst stage. These cysts then settle to the ground as the snow melts to remain inactive until the next spring. Researchers have found that the algae have unique fatty acids that enable them to remain flexible in the cold environment in which they spend their entire active lives.

Snow algae have been uniquely designed by God to make a living in some of the lowest temperatures on Earth. If our Creator can find a way to enable snow algae to flourish in these conditions, He can find a way to help us out of the most impossible circumstances.

Prayer: Lord, help me to trust your wisdom to solve my problems. Amen.

Ref: *Science News*, 5/20/00, pp. 328-330, "Red Snow, Green Snow."

Can Storms on the Sun Kill Here on Earth?

Luke 12:55
"And when you see the south wind blow, ye say, 'There will be heat; and it cometh to pass."

Man has always been fascinated with the weather. Some of the effects of weather on health are obvious. Death rates rise in areas experiencing extreme heat waves or severe air pollution. Researchers also agree that too little sunlight can result in depression in many people. And a 1997 study by the University of Massachusetts Medical School found a correlation between falling barometric pressure and the onset of labor for pregnant women.

The most dramatic weather link, however, involves geomagnetic storms. These storms are caused when storms on the sun compress or decompress the Earth's magnetic field. Researchers noticed increased numbers of babies suddenly dying in their sleep during geomagnetic storms. This is known as crib-death. It was known that crib-death victims had low levels of melatonin before they died. Melatonin controls the amount of nitrous oxide, an essential compound for breathing, present in our blood. Researchers theorized that the magnetic effects of a geomagnetic storm could decrease melatonin levels during sleep and therefore result in breathing problems. They tested this theory by subjecting baby rats to low intensity magnetic fields. The baby rats did indeed show a decrease in melatonin before they stopped breathing. This evidence indicates that geomagnetic storms are likely to trigger crib deaths or SIDS (Sudden Instant Death Syndrome). It is hoped this research will come up with a way to prevent these tragic deaths.

Although the perfect world that God created was ruined by sin, we are told that He will one day redeem it.

Prayer: Lord, prosper all research that improves our lives here in this life. Amen.

Ref: *Discover*, 6/00, pp. 78-81, "Is the Weather Driving You Crazy?"

World's Most Unusual Lake Challenges Evolution

1 Peter 4:19

"Wherefore let them that suffer according to the will of God commit the keeping of their souls to him in well doing, as unto a faithful Creator."

Siberia's Lake Baikal provides many mysteries for those who believe that naturalistic evolution can account for today's biological world. Many of the unique plants and animals that call this fresh water lake home are similar to those found only hundreds of miles away in marine environments.

Lake Baikal is the deepest lake in the world—5,371 feet deep at its deepest point. Located in central Siberia, the lake has a huge variety of life. It boasts 1,550 varieties of animals and 1,085 varieties of plants. Of these, over 1,000 species are found no where else on Earth. Seals are among the animals you would not expect to find in a freshwater environment nearly 500 miles from the sea. Lake Baikal also supports other sea creatures you would not expect to find in a fresh water lake. Hydrothermal vents were recently discovered in the depths of the north end of the lake. These are the only known hydrothermal vents in any fresh water lake. Like their marine cousins, these vents support a rich variety of sponges, transparent shrimp, bacterial mats and fish. Again, some of these creatures are found no where else on Earth.

The questions posed by the plants and animals of Lake Baikal are troublesome for evolutionists. How did so many marine species successfully evolve to thrive in fresh water? Why are so many of these unique to a lake that is so much like an inland ocean? Did evolution know that Lake Baikal was the most ocean-like lake in the world? The plants and animals are better accounted for by the creative powers of the Creator of all that exists.

Prayer: Dear Father, help me to trust your creative wisdom in my life. Amen.

Ref: *Science Frontiers*, p. 133, "Baikal's Deep Secrets."

248

Bees Have a Hot Line of Defense

Proverbs 18:14
"The spirit of a man will sustain his infirmity; but a wounded spirit who can bear?"

When bees are threatened by some diseases or predators, they act like sick children—they get a fever. More specifically, they give their nest a fever, sometimes raising its temperature to levels nearly fatal to themselves.

Bees are cold-blooded, but they can generate heat by flexing their flight muscles while holding their wings still. It has long been known that bees can generate temperatures as high as 96.8° (F) to keep the nursery nice and cozy. Researchers have found that this ability is used to cure a sick hive. One of the more dangerous threats to the bees' nest is the chalkboard fungus. Researchers noticed that bees raised the temperature of the nursery when threatened with the fungus. They then tested the effect of the temperature increase by introducing chalkboard fungus spores into three experimental hives. Even before the larvae showed symptoms of infection, bees raised the temperature in all three hives. While several larvae mummified in one hive, no larvae in the other two hives showed infection. Other researchers have also found that bees use the same strategy on giant hornets. The bees cannot sting through the hornet's tough armor. So, when invaded, bees raise the hive's temperature to over 116°(F). That's enough to kill the hornet, but one degree hotter would be fatal to the bees.

God has given His creatures important abilities to gain and keep physical health. But when it comes to spiritual health, we must rely completely on what God's Son, Jesus Christ, has done to give us that spiritual health.

Prayer: Lord, I thank you for giving me spiritual health through forgiveness. Amen.

Ref: *Science News*, 5/27/00. p. 341, "The whole beehive gets a fever."

Speed of Light Reveals the Unlimited Mind of God

Romans 11:34
"For who hath known the mind of the Lord? or who hath been his counsellor?"

All the textbooks inform us that the speed of light is about 186,000 miles per second. This is the absolute speed limit of the universe, according to Einstein's theory of relativity, we have always been told. Should anything travel faster than that, Einstein's theory falls.

While the certainty of the speed of light as an absolute limit has provided comfort for many, new research may point to a universe that is stranger than we can imagine. It all began when an Italian researcher showed that microwaves could travel through the air at speeds slightly faster than light. Then an American researcher decided to see what would happen to the speed of light in a special chamber. He built a clear chamber filled with cesium gas. When a pulse of light was sent through it, the light went through the chamber at 300 times the speed of light. This is so fast that the main pulse of light exited the chamber before it went in! This strange effect is a result of the unique wave action of light. For this reason, researchers say that Einstein's theory remains viable. Still this finding has led to a debate among physicists over whether a message could be sent faster than the speed of light into the past.

This research should remind us that the universe is the creation of an unlimited, almighty God. While scientific research can bring many benefits, it cannot be too dogmatic about its interpretation of the universe. Only the Bible, which is the very Word of the unlimited, almighty God, offers absolute knowledge.

> **Prayer: I thank You Lord, that you are unlimited, especially in your love for me. Amen.**

Ref: *The Denver Post*, 5/30/00, p. 2A, "Raising the speed of light."

Structure of Atoms

Hebrews 11:3
"Though faith we understand that the worlds were framed by the word of God, so that things which are seen were not made of things which do appear."

Let's go back to the very first moment that the supposed Big Bang took place. Evolutionists say that at that very first moment, everything was up to chance. The size and charge of electrons, protons, the structure of atoms or whether they would even exist could have been anything.

Now let's move forward to what we actually know about the various atoms and their structure. Life as we know it is based on the carbon atom. Its structure makes it the only atom with almost unlimited ability to share pairs of electrons with other atoms. This makes possible the rich range of biological molecules needed for life. No other atom can do carbon's job. The oxygen atom's structure causes it to bind together in pairs. This type of bonding leaves unpaired electrons that allow oxygen to bind with iron. This feature makes hemoglobin capable of carrying oxygen in the blood. There are several other atoms that could replace iron in hemoglobin, but they would hold the oxygen either too tightly or too loosely. So there are no substitutes for iron. Likewise, the zinc atom is the only atom that can allow proteins do the crucial job of identifying their own unique DNA sites.

The precise structure of atoms was clearly not the result of chance. Each was carefully designed by the Creator to support the life He would form only days after He made the atoms.

Prayer: I thank you, Lord, not just for the gift of life, but for eternal life, too. Amen.

Ref: *Impact* (ICR), 6/99, "Basic Chemistry: A Testament of Creation."

Faster than the Speed of Light?

Psalm 40:5a
"Many, O Lord my God, are thy wonderful works which thou hast done; and thy thoughts which are to us-ward: they cannot be reckoned up in order unto thee..."

Often the most productive science is done when scientists, as one founder of modern science put it, "Think God's thoughts after Him." For example, while antibiotics have saved millions of lives, they have proven to be only a temporary solution. In addition, there are some bacterial infections that move too fast for antibiotics to be effective.

Vibrio vulnificus is one such bacterium. It is typically found in oysters. The bacterium has little effect on healthy people, but those whom it infects are most likely to be dead within 24 hours. How do we combat such a fast-working deadly bacterium by thinking God's thoughts after Him? By finding something that is even faster at infecting the bacterium and killing it. That's exactly what medical researchers have done. They have found a virus that infects the bacteria, rapidly reproducing inside the bacterium until it causes the bacterium to burst. Once the bacteria are all dead, the virus can no longer reproduce. Researchers infected eight mice with *Vibrio vulnificus*. If they had not at the same time infected the mice with the virus, all would have been dead within 18 hours. As a result, five of the mice never became ill. This approach is also being successfully used to treat bacteria that cause skin infections.

The successes that come from thinking God's thoughts after Him demonstrate that God is the Author of science and all that it studies.

Prayer: Dear Father, help me to think more as you think. In Jesus' name. Amen.

Ref: *Science News*, 6/3/00, p. 358, "Viruses that slay bacteria draw new interest."

Screen Actor Octopus

Psalm 81:15
"The haters of the LORD should have submitted themselves unto him; but their time should have endured for ever."

The animal we feature on today's Creation Moments is so new to science it hasn't even been classified according to genus as this is written. Yet the mimic octopus is the talk of the scientific world.

Found in the Indo-Pacific area, the mimic octopus has proportionately longer arms than other octopuses. The mimic octopus also has a very narrow waist that aids in its disguises. When approached, one mimic octopus mimicked a flounder, flattening its body and pointed its legs in one direction. Then it began to swim with undulating motion over the bottom, just as a flounder would. Mimic octopuses have also been seen to mimic starfish, jellyfish, giant crabs, lionfish, seahorses, anemones, stingrays and other creatures. One researcher was studying what he was convinced was a new species of mantis shrimp. It turned out to be a mimic octopus. If the possible threat is not fooled, a mimic octopus will morph through a variety of animals. Octopuses are generally seen as intelligent animals. However, the ability of the mimic octopus to mimic both the form and behavior of a range of animals puzzles scientists.

According to evolutionary theory, such high intelligence evolves only in social animals, and mimic octopuses are solitary creatures. So where did such intelligence come from? We can easily explain that its intelligent Creator gave it the intelligence it needed to fool predators for its own survival.

Prayer: Lord, help me always to be honest, especially with you. Amen.

Ref: *Asian Diver*, (on-line), 5/22/00.

Race for the Pyramids

2 Chronicles 12:2a, 3b
"And it came to pass, that in the fifth year of king Rehoboam Shishak king of Egypt came up against Jerusalem ... and people were without number that came with him ... and the Ethiopians."

While most people know that several ancient cultures built pyramids, which culture built the most? Most people would answer "Egypt" but they would be wrong. However, they did inspire the culture that built the most pyramids.

At its height, the Egyptian empire included many cultures, reaching into the Near East as well as into Africa. Nubia, just to the south of Egypt, was a proud member of the Egyptian empire and adopted many Egyptian customs. Sometimes referred to as Ethiopians in some Bible translations, they became so much a part of the Egyptian empire that they even provided troops to the Egyptian army, according to the Bible. Nubia is also called "Cush" in the Bible. Among the Egyptian customs they adopted was burying their kings in pyramids. After mummification, dead royalty were placed in their tombs, which were decorated and filled with supplies thought to be needed in the next life. The base of these pyramids also included a chapel for those who wished to continue to worship the dead king. While none of their pyramids was as large as the great Egyptian pyramids, over the centuries they built 223 pyramids, far more than are found in Egypt!

The Bible's mention of these people serving in the Egyptian army rings true to history. The Nubians were proud members of the Egyptian empire and continued to enjoy an alliance with Egypt long after the empire ended.

Prayer: I thank you, Lord, that your Word can be trusted in all that it says. Amen.

Ref: *Oriental Institute* (on-line), "The Kingdom of Kush: Napata."

Babies Baffle Evolution

Genesis 11:1
"And the whole earth was of one language, and of one speech."

The Bible teaches that until several centuries after the Flood all people on Earth spoke the same language. The fact that many supposedly unrelated languages have similar sounding words with the same or similar meanings supports this teaching. Some language researchers say that this is simply due to chance. Now some language researchers say that they have developed another line of research that supports the belief that all humans once spoke the same language.

Language researchers have spent years listening to babies in many different language groups babble. Babies younger than four months make a variety of sounds as they learn to use their voice. Speaking is a highly complex activity, requiring the coordination of 70 different muscles as well as several different body parts. By seven to ten months of age babies typically begin making sounds that alternate vowels and consonants. Researchers found that there are three distinct patterns of alternating sounds that are universal among babies in English-speaking households. Then they found that these three patterns are also common among infants from a wide range of language groups around the world. Researchers comparing infants from varying language groups then identified a fourth pattern among all groups. These findings have been interpreted as independent evidence that all people once spoke one language.

We accept what the Bible says by faith, not because it has been proven. But when Scripture's truths are supported by science the Bible's integrity is upheld before skeptics.

Prayer: Father, help me use the gift of language to tell others about you. Amen.

Ref: *Science News,* 5/27/00, pp. 344-346, "Building Blocks of Talk.")

Boys Training to be Girls?

Genesis 1:27
"So God created man in his own image, in the image of God created he him; male and female created he them."

Do men and women differ from each other because God gave them different natures, or because their environment caused them to be men or women? In short, is it our nature (our genes) or our nurture (our environment) that determines whether we become men or women?

Perhaps a pair of Canadian twin boys can answer the question. When they were a few months old, their parents decided to have them circumcised. After the doctor made a tragic error with one of the boys, the distraught parents did not know what to do. Not long after that, they saw a program on television about the effects of nature and nurture. During the course of the program, a doctor claimed that a surgically altered boy who was raised as a girl would grow into a woman. The parents contacted him. He assured the parents that he could help them raise their son as a healthy, happy girl. However, as the years went by, their new "daughter" grew to behave more like a boy than a girl. Finally, as a young adult he reclaimed his identity as a male. Today he is married and the stepfather of three children. His experience has led him to call the idea that it is the way children are raised that determines whether they become men or women, "Plainly ignorant." The doctor who encouraged his parents to raise him as a girl has since been removed from his gender studies clinic.

We are made male or female by God, Who gives each of us the unique characteristics we need for our unique roles as parents. No amount of denial of this truth can change the truth of what God creates.

Prayer: Thank you, Dear Father, for making me what I am. In Jesus' Name. Amen.

Ref: *The Toronto Star,* 2/6/00, p. D13, "From boy to girl to boy."

Two Noses Are Better than One

Psalm 115:4,6
"Their idols are silver and gold, the work of men's hands... They have ears, but they hear not hear; noses have they, but they smell not..."

Did you know that most animals and humans have two noses? Most animals and humans have two very separate systems for detecting scents that work very differently from one another. These two sensory systems even detect very different types of scents.

We are all familiar with the smell of dinner cooking. However, scientists are making some surprising discoveries about our other system for detecting scents. This second system uses an organ called the vomeronasal organ, or VNO, in the nose. Rather than smelling dinner, the VNO detects pheromones. Among animals pheromones are important for mating behavior. Mouse studies show that the VNO is wired into the brain with its own set of neurons. The research shows that mouse VNO neurons are up to 10,000 times more sensitive to pheromones than nasal neurons are to other scents. This amazing sensitivity rivals that of insects' abilities to detect pheromones. While your nose uses many receptors to detect scents, your VNO apparently works quite differently.

The ability to detect scents is amazing enough, and could hardly have developed by evolution. That we have two very different systems that fulfill very different purposes is a tribute to God's creativity and shows that evolutionary thinking is nothing but a modern idol.

Prayer: I thank you, Dear Father, that you love us because you have made us. Amen.

Ref: *Science News*, 6/17/00, p. 390, "Mice have a sharp nose for pheromones."

Another 19th Century Creationist

Proverbs 2:6
"For the LORD giveth wisdom: out of his mouth cometh knowledge and understanding...."

John Stevens Henslow was a highly respected scientist of the 19th century. He taught both botany and mineralogy at England's Cambridge University. In addition to being a scientist and professor, he was also a devout Christian and ordained into the Anglican clergy.

Charles Darwin was an undergraduate at Cambridge studying for the ministry. However, after his first year of study he found he had no interest in religion. Because of his enthusiasm for his subjects, John Henslow was one of the most popular professors at Cambridge. That probably contributed to the friendship Charles Darwin developed with him. Darwin's friendship with Henslow was also an opportunity to see living Christianity in action. However, while Darwin learned a great deal about botany from Henslow, he rejected Henslow's Christian faith. In 1831, when Darwin received his B.A. degree, Henslow recommended Darwin be an unpaid naturalist on the H. M. S. *Beagle*. It was during that five-year voyage on the *Beagle* that Darwin began to formalize his theory of evolution. Darwin's book "Origin of the Species" was finalized and published almost 30 years after the voyage. The aging Henslow publicly expressed his opposition to Darwin's theory, stating that, "Darwin attempts more than is granted to man...."

Henslow was only one of many outstanding 19th century scientists who rejected Darwin's theory. He understood that anyone who promotes so-called knowledge that contradicts Scripture dangerously places himself above God.

Prayer: Lord, expose those who would contradict your truth and protect us. Amen.

Ref: *Creation Matters* (CRS), 3-4/00, p. 2, "John Stevens Henslow (1796-1861)".

Get the Scoop on Nuts!

Genesis 43:11
"And their father Israel said unto them, `If it must be so, do this; take of the best fruits in the land ... nuts and almonds.'"

New research is revealing the health benefits of eating nuts. While eating peanuts has been shown to reduce heart disease risks by lowering triglycerides, tree nuts offer many more benefits.

People often worry about the fat in nuts. However, most of that fat is the good kind that protects hearts, while at the same time satisfying the body's need for fat. The result is that nut eaters can lose weight and keep it off. In one Canadian study, those trying to lose weight while satisfying their fat needs with nuts lost an average of nine pounds over 18 months. Over the same trial period those trying to lose weight with a low-fat diet actually gained an average of six pounds. Another study showed that people who ate two ounces of nuts four or five times a week lowered their risk of heart attack by 50 percent. Yet another study showed that women who ate only five ounces of nuts per week were one-third less likely to have a heart attack than those who ate nuts infrequently. Several studies have shown that almonds, macadamia nuts, or walnuts lowered high cholesterol by ten percent. Other studies have shown that nut eating can lower the risk of prostate cancer by 30 percent and may lower the risk of colon and stomach cancers.

Nuts are a great gift from God and truly are among the best fruits of the land. Chance would not have evolved all these health benefits in a food clearly designed for us to eat. A loving Creator made them for our good.

Prayer: Thank you, Dear Father, for all the good gifts of the earth. Amen.

Ref: *USA Weekend,* 6/23-25/00, p. 6, "Nuts, your new superfood."

Contact us to order:

- "Letting God Create Your Day" book.

- "Creation Moments" on cassette or CD
  (each containing 30 programs).

- Ministry information and a resource catalog.

Creation Moments, Inc.
P. O. Box 260
Zimmerman, MN 55398
1-800-422-4253
www.creationmoments.com